Bad Jews

Also by Emily Tamkin

The Influence of Soros

Bad Jews

A History of American Jewish Politics and Identities

Emily Tamkin

HARPER
An Imprint of HarperCollins*Publishers*

HarperCollins books may be purchased for educational, business, or sales promotional use. For information, please email the Special Markets De-partment at SPsales@harpercollins.com.

FIRST EDITION

Library of Congress Cataloging-in-Publication Data has been applied for.

ISBN 978-0-06-307401-9

22 23 24 25 26   LSC   10 9 8 7 6 5 4 3 2 1

For my parents, who raised me to be so proud to be Jewish, and for Neil, who is not Jewish but with whom I am so proud to build a Jewish home, and in loving memory of my grandparents, Alvin and Jacqueline Tamkin. I know you would have disagreed with so much in this book. I hope it would have made you proud, too.

There is no other people like the Jewish people, that talks about itself so much, but knows itself so little.

—S. ANSKY

Everyone will sit under their own vine and under their own fig tree, and no one will make them afraid.

—THE BOOK OF MICAH

DAPHNA: *Melody, I can't get a tattoo. I'm Jewish? It's against Jewish law.*
MELODY: *I know a Jewish person with tattoos.*
DAPHNA: *Well they're wrong.*

—JOSHUA HARMON, *BAD JEWS*

CONTENTS

Bad Jews

INTRODUCTION

WHAT DOES IT MEAN to be a Bad Jew?

The question is not a new one. Many Jews—at least many Jews I have known and loved or known and not loved—throw the term at and against one another. You can be called a Bad Jew if you don't keep kosher; if you don't go to temple often enough; if you don't go to or send your children to Hebrew school; if you enjoy Christmas music; if you date or marry a non-Jewish person; if you don't call your mother often enough. The list goes on.

In the course of writing this book, when I asked American Jews "what comes to mind when you hear 'Bad Jew'?" I got answers varying from someone who casts Jewishness aside and sees it as having no value to someone who rejects the connection between Jewishness and Zionism. I heard that a Bad Jew is someone who fails to hold Jews and Judaism to a high moral standard. Stephen Miller, for example, the former Trump official who crafted and pushed for the implementation of much of the administration's far-right immigration policy, was considered a Bad Jew by many who felt he wasn't living up to Jewish values. Others said a Bad Jew is someone whose conception of Judaism doesn't have applications to the wider world; someone who clings to outdated notions of how to talk about other Jews. Rebecca King, a Los Angeles–based chef who runs The Bad Jew, a restaurant that sells "pork-strami," said that the phrase makes her think of Jewish kids running around who just want to eat bacon.[1] The most common answer I got was, "When I think of a 'Bad Jew,' I think of myself."

For years, if asked this question, I, too, would have given that answer: that it was me, that I was what I thought of as a Bad Jew.

The issue of what it means, or doesn't, to be a Good Jew or a Bad Jew is particularly fraught at this moment in US history. American Jewish institutions are saying one thing; American Jews at large say another. The demographics of American Jews are changing as are American Jewish politics.

And so there is, at the time of writing, a debate over who gets to speak for American Jews and who gets to claim American Jewishness. Is there a single American Jewish community? Or many American Jewish communities with competing values? What does it actually mean to be an American Jew? And if we don't know, why does the label bring with it so many unspoken assumptions about how a person is and should be?

The concept of "American Jewishness" is one of identity ("I am an American Jew") and belonging ("I am accepted as an American Jew by other American Jews"). These two concepts can run into each other. What happens when one, or many, identify as American Jews but in a way that is unfamiliar or even unacceptable to others who also identify as American Jews? There are several million people who identify as American Jews. But that doesn't mean they all identify with one another. And it doesn't mean they all accept one another. There may be people who will read this book and decide that I am not really an American Jew. What do I do with that? What do they?

The conflict between American Jews was salient in the years that I was thinking of, and finally actually writing, this book. As with much in the United States today, this was at least in part because of Donald Trump.

Over the course of his administration, President Donald Trump worked closely with Israeli prime minister Benjamin Netanyahu, stopping aid to Palestinians and moving the US embassy in Israel from Tel Aviv to Jerusalem.[2] He also described his opponent in the US 2020 presidential election, Joe Biden, as a "servant of the globalists,"[3] a term the Anti-Defamation League (ADL) has described as being "used as an ethnic smear and long-running conspiracy theory about Jewish populations not being loyal to the countries they live in and cooperating through secret international alliances."[4] Right-wing Jews and their allies, including certain Evangelical Christians, who were some of Trump's most ardent supporters, argued that Trump was a great friend to Israel and the Jewish people.[5] Partisan conservative Jewish organizations, like the Republican Jewish Coalition, dismissed accusations of antisemitism against Trump[6] and other Republican politicians who dabbled in antisemitic tropes—an accusation that a Jewish billionaire was buying an election, or a digital ad altered to make a Jewish candidate's nose look larger, to name two.[7] But mainstream, ostensibly nonpartisan Jewish organizations also drew criticism for pulling punches, celebrating the Trump administration's Israel policy. Some progressives argued that these groups were at least as loudly repudiating the Left as they were censuring Trump, the most powerful person in the country.[8]

Meanwhile, liberal Jews—the majority of American Jewish voters—noted that American Jews have a range of opinions on Israel, and argued that Trump was enabling antisemitism at home, which was decidedly bad for Jews.[9]

Some leftist Jews, for their part, argued that the term "antisemitism" had been weaponized by the political Right and warped beyond recognition,[10] and pressed politicians not on their support for Israel, but for Palestinian rights.[11]

The conversation of what role Israel should play in American

Jewish politics, and what role American Jews should play in Israeli politics, was also changing. Some Jewish groups present themselves as though they are speaking on behalf of all Jews, or at least in favor of what is best for and in the interest of American Jews. But who, here, is the Bad Jew? Is it the American Jew who speaks against Israeli policy? The American Jew who remains muted on antisemitism because of Israel? The American Jew who purports to know what is best for other American Jews?

The shift both seems sudden and is years in the making. A decade ago, in 2011, Ben Shapiro, a conservative political commentator, wrote a tweet that read, "The Jewish people has always been plagued by Bad Jews, who undermine it from within. In America, those Bad Jews largely vote Democrat."[12] That same year, he penned an article decrying JINOs, or Jews in Name Only. Seventy-eight percent of American Jews may have voted for Obama in 2008,[13] but, to Shapiro, "the Jews who vote for Obama are, by and large, Jews In Name Only (JINOs). They eat bagels and lox; they watch 'Schindler's List'; they visit temple on Yom Kippur—sometimes. But they do not care about Israel. Or if they do, they care about it less than abortion, gay marriage and global warming."[14]

Shapiro was taking the trope of the Bad Jew and mapping it onto the American political spectrum. In this telling, the right kind of Jew regularly attends temple and cares first and foremost about Israel. Since the vast majority of American Jews vote for Democrats and have for decades, he was denouncing the vast majority of American Jews as Bad Jews. The Trump years only made this paradox more pronounced—most American Jews did not support him for president, but some of the loudest and most prominent voices were claiming he was good for the Jews.

Over the past several years, the idea of what role American Jews could play as politicians and political actors has changed, too. Two

Jewish men—Michael Bloomberg, the billionaire ex-mayor of New York City, and Bernie Sanders, the Brooklyn-born Socialist—ran for president, two different paradigms of American Jewish identity, the financial powerhouse and the rabble-rouser, standing on the primary debate stage.[15] Each performed American Jewishness differently. And it wasn't only the candidates who were changing American Jews' position in political discourse. Even outside of issues relating to Israel, some American Jews protested Trump, citing *tikkun olam*, the Jewish concept that means taking responsibility for repairing the world, in doing so[16] (though actually the concept comes from a prayer about removing non-Jewish idols from Earth).[17] Other American Jews went so far as to work in the Trump administration: the aforementioned Stephen Miller was the architect of some of the administration's harshest, most discriminatory policies.[18] Some American Jews argued that Jews should work with other minorities and oppressed people for universal justice; others, that Jews specifically are overlooked as a minority group, or even that certain approaches toward racial justice, or academic framings like Critical Race Theory, hurt Jews.[19]

I would argue that the fact that we are in a time of change and conflict and challenge has thrown many American Jews off-balance. Things are not as they were. But that, in turn, means there is an opportunity to think about what things could be.

American Jews debate what it means to be Jewish in the United States. But we also disagree on what constitutes a threat to American Jews.

Antisemitism, sometimes called the oldest hatred, is defined as the hatred of or prejudice against Jewish people. I personally believe, looking historically and presently at antisemitic conspiracies, be they that Jews caused the Black Plague or Jews control the world's economy or Jews are orchestrating protests on the streets of the United States, that it would perhaps be more accurate to say that antisemitism is the conviction that Jews are forever foreign or alien to whatever population they happen to be in, and often have designs on corrupting that population.

The term itself goes back to the nineteenth century,[20] but hatred of Jews as a people dates back hundreds of years.[21] Many of the tropes that we, today, think of as antisemitic—that Jews have outsize influence over society or even control the world, or that Jews are greedy and obsessed with money and run financial systems, or that Jews are trying to bring about the downfall of a country or nation or (before the advent of nations) communities—go back hundreds of years, too. Take, for example, the Spanish Inquisition, which began in the fifteenth century and resulted in the expulsion of Jews. And though the intensity of antisemitism has ebbed and flowed at different times and in different places, it has never fully abated. And indeed, Jewish immigration to the United States peaked as a result of antisemitism in Europe in the late nineteenth and early twentieth centuries.

Jews didn't escape antisemitism by coming to the United States, however. Immigration was capped because of antisemitism, which intensified and reached a fever pitch during World War II. Then, in the postwar period, antisemitism in the United States, or at least expressions of it, decreased.[22]

In recent years, however, antisemitism is once again perceived not as latent but explicit.

"You're seeing it happen here," my father told me when I interviewed him for this book. He grew up in a heavily Jewish community, but nevertheless can recount antisemitic incidents from his childhood. But it is only recently, he said, that he's observed "large segments of the American population are starting to blame Jews for their problems."[23]

"When I was in college in the 1980s and went to law school, it looked like virulent antisemitism was a thing of the past in America," Representative Jamie Raskin, a congressman from Maryland, told me when I interviewed him for a piece on antisemitism in 2020. "It was not a constant in people's political lives. And now suddenly antisemitism—and racism—are a matter of daily conflict and contest."[24]

My father and Congressman Raskin are not alone in thinking so. The ADL found that antisemitic incidents hit an all-time high in 2019 (it began tracking in 1979).[25] In its 2020 survey of American Jews, the American Jewish Committee found that, while most American Jews had not experienced antisemitism directed at themselves personally, 88 percent of American Jews felt antisemitism was "a problem." Eighty-two percent felt it had increased over the past five years.[26]

But here, too, there is little consensus. While the same American Jewish Committee survey found that 69 percent of American Jews believed Republicans held a lot or some antisemitic beliefs, compared to just 37 percent of American Jews who believed the same about Democrats, there are, nevertheless, some who insist that the more insidious antisemitism comes from the political, and in particular the anti-Zionist, left.[27] Whether anti-Zionism should count as antisemitism is itself a matter of debate.

Relatedly, after Joe Biden's election to the presidency in 2020, American Jewish institutions debated whether to adopt and encourage as policy a definition of antisemitism that was published in 2016 by the International Holocaust Remembrance Alliance. Some worried, given the explanatory examples provided along with the definition, whether it could easily be interpreted to silence criticism of Israel by conflating the country with the Jewish people.[28] Some, like historian David Engel, argue that the term "antisemitism" is altogether flawed and arbitrary.[29]

American Jews are not the only ones who have tried to define antisemitism in recent years. In his last weeks as secretary of state, Mike Pompeo declared that anti-Zionism is antisemitism. In his 2021 confirmation hearing, Attorney General Merrick Garland, whose family came to this country to escape antisemitism, was asked by Senator Mike Lee of Utah, who is not Jewish, about comments Garland's future deputies had made about Israel. "Senator," Garland replied, "I'm a pretty good judge of what an antisemite is."[30]

There is, then, little consensus on what it means to be Jewish not only in the active sense—what a person who is Jewish should do and say and support—but also in the sense of what American Jews should be fearful of. We are a minority in this country, which can bring with it risk and fear. But we don't agree on what we should be afraid of.

As specific as this moment is, though, it is also but the latest in a series of such moments in American Jewish history, which is full of discussions and debates and hand-wringing over who is Jewish, and how to be Jewish, and what it means to be Jewish.

There have been times in US history when American Jews have had harm done to them, and there have been times in US history when American Jews have done harm to others. There have been times in US history when some American Jews have harmed other American Jews. And there have been times—including right now—when American Jews have disagreed on which American Jews are the victims and which the villains. This complex and complicated history is one that I hope to unpack, and in which I will aim to situate present debates.

Why bother with the history? Why would someone like me—a journalist of present-day society and politics—wade into it?

The first reason is that the contemporary insistence that there was and is a true way to be Jewish has consequences. The insistence that intermarried Jews are less serious about Judaism and somehow not fully capable of passing Jewish values on to their children, for example, affects how their children are welcomed, or not, into Jewish spaces. The conflation of Jewish with a certain physical appearance leads Jews of color to be treated as outsiders in their own synagogues. The argument that real Jews only have certain politics threatens to push out Jews who think, speak, and act differently. We should try to both name and understand what is happening and to put today's debates on Jewishness in their appropriate historical context.

And the second reason is that being Jewish—like being anything—can mean many things. I hope that this book will be useful for American Jews, or indeed anyone interested in thinking about any kind of identity, to consider the strengths and weaknesses of labels. How limiting they can be. And, at the same time, how much room there is in them, how much elasticity to stretch and shape them into what you want them to be.

In a 2020 interview with Dr. Josh Perelman, chief curator at the National Museum of American Jewish History, I said that Jews were neither victims nor villains of American history. He gently corrected me.

It isn't that American Jews were neither victims nor villains, but that they—we—have been both victims and villains. The challenge, he said, is to accept that truth, explore its meanings, and unpack it.[31]

Bad Jews is a roughly hundred-year history of Jewish American politics, culture, identities, and arguments.

This book begins as mass Jewish immigration to the United States ends. There have been Jews in the United States for as long as there has been a United States. My reason for choosing the legal cutoff of mass Jewish migration as a starting point is that, from the 1920s on, with the stream of Jewish immigrants reduced to a trickle, the Jewish population in the United States became less of an immigrant population. It also became more of a population that was left to grapple with, and disagree about, what it meant to be in and of America.

The book then moves both historically and thematically to deal with evolving and conflicting Jewish positions on assimilation; race; Zionism and Israel; affluence, philanthropy, finance, and poverty; politics; and social justice. It will, in the process, attempt to tackle Jewish stereotypes, examining how they came to be and what makes them so dangerous, not only to American Jews, but to America. But more than that, it will try to go beyond the stereotypes and grapple with the various ways in which American Jews have tried to figure out how to position themselves on these issues and in the United States.

While reporting this book, I interviewed roughly 150 people, the vast majority of whom are Jewish. The book was written during the global coronavirus pandemic, and so was mostly reported out by phone or over WhatsApp or Zoom from my home in Washington, DC, as I sat at my counter or on the couch and apologized for the sound of my dog shaking her collar while I asked people what being Jewish meant to them. Parts were also reported in New York City, Tel Aviv, and Jerusalem (indeed, part of it was written while quarantining for seven days in a Tel Aviv hotel). Though Israel-Palestine looms large in the American Jewish imagination and consciousness, and so features in this book, the bulk of it is set in the United States and is about American Jews. For this, I spoke not only to historians and academics and journalists and policy makers and politicians and significant figures in the Jewish American community, but also, mostly, to American Jews who are not any of those things. American Jewish history is a living history, I said when I put out the call, and so I wanted to speak with living American Jews.

One woman, Ruth Boehl, emailed me ahead of our interview. She wanted to be sure, before I took the time to speak to her, that she was the right kind of Jew. I assured her that she was. It became clear, over the course of our interview, that the question came from a lifetime of being treated like she was not.

The reason she filled out my little form for potential interviewees, she told me, was because of the book's title.

"I think of this all the time," she said. "I think of the inverse. What is a Good Jew?" she asked, half kidding. "Will someone explain to me what a Good Jew is? Someone, please tell me."[32]

This book does not contain an answer to that question. Rather, this book will try to use American Jewish history to demonstrate that the concept of "Bad Jew"—much like that of "Good Jew"—is, like almost everything else in American Jewish history, more contested than defined, and that there are more meaningful conversations about American Jewishness to be had instead.

I hesitated, at first, to write this book. I wondered whether I had the right to write a book on American Jewish history. My mother's family is not Jewish; she converted before I was born. I went to a Jewish preschool, but that was the end of my formal Jewish education. My family moved back to the United States from Canada when I was eight to an overwhelmingly non-Jewish town on Long Island. My father, who grew up in a predominantly Jewish town in Massachusetts, did not want us to think that the world was all Jewish. New in a town that turned out to be almost cartoonishly antisemitic, I would have had to go to extra Hebrew school classes to catch up in time for my bat mitzvah, and my parents decided against that, which is a long way of saying that I did not go to Hebrew school, did not learn the language, and never had a bat mitzvah. We grew up celebrating the Jewish holidays, but mine was not a religious upbringing. I studied Russia and Eastern Europe, not Jewish or Middle Eastern Studies, in college and graduate school. I did not actually go to Israel until I started working on this book. I eat shellfish. While conducting interviews for this book, a Jewish person referred to me as "an outskirts Jew."

I only ever dated one Jewish person; after we broke up, he messaged me on Instagram on the fifth night of Chanukah to let me know that I had lit my candles backward. My husband is not Jewish, and though we did have a Jewish wedding ceremony, it was held not only on Shabbat, but on Simchat Torah; finding a time and a day that we could have our immediate families safely present during a pandemic was more important to us than respecting the Jewish calendar (happily, the rabbi who performed the ceremony agreed). During the pandemic, I started learning Yiddish. We joined a Reform synagogue, a community and, one day, hopefully, a place where our future children can learn about being Jewish. I felt like a person playing at being Jewish while filling out the membership forms, just as I sometimes do when we light our Shabbat candles every Friday night.

I worried, in other words, that I was not Jewish enough, or not the right kind of Jewish, to write a book on American Jewish history.

What I realized, though, was that I was not only holding myself to a different standard than that to which I would hold anyone else, but a standard that I would actively argue against if I heard it applied to anyone other than myself. And I realized this, too: No one person is an authority on being an American Jew. We can and should try to acknowledge and include the many ways of being Jewish in America. I, for example, have tried to include the histories and perspectives of traditional Orthodox, Sephardic, and Mizrahi Jews, as well as those of Jews of color and Jews who converted, though I am none of those things. But as I am none of those things, this book is, inevitably, different than it would be if it were written by someone who is not Ashkenazi, or who grew up in the Conservative Jewish tradition, or who themselves converted to Judaism. So often in writing this book, I realized that a given section could have been its own tome, or that I could have included a completely separate set of examples, incidents, and anecdotes and still filled a whole book. Like any work on identity, this one is imperfect and incomplete.

Still, it is my best attempt to wrestle with what I believe to be the one truth of American Jewish identity: it can never be pinned down. There are, at any given time, dominating narratives about what it means to be Jewish in America, but they are just that: narratives, and counternarratives, and counternarratives to the counternarratives. Stories we tell ourselves.

Chapter 1

Foreign Jews

IN 1912, SOLOMON CAME to the United States. His family settled in Boston, Massachusetts, one of the primary sites for Jewish immigrants, though not the main one, a distinction held by New York City. He worked a variety of odd jobs and then made his way to selling ice. "*Everybody* bought ice," he later bragged to his grandchildren in a recorded conversation between bites of corned beef, chicken, and bread rolls. He made money. But it was "too much hard work" to do it by himself, so he turned it into a business. In time, he started selling coal, too.

He soon brought his parents, Benjamin and Ida, over to the United States, though not his brother and sister, who died back in Korets (*Koritz*, in Solomon's native Yiddish), which was then part of the Russian Empire and is now in Ukraine. Years later, when his children asked him about the place he was born, Solomon replied, "If it had been any good, we would not have left." He would eventually marry Fannie Chasin, whom he met in a grocery store, trying to sell her coal. Fannie was also originally from the Russian Empire. She was a stocky woman ("like a tomato," Solomon recalled) and a natural-born neighborhood busybody.[1] They had left the intense and violent antisemitism of the Russian Empire and could now—*would* now—make a life for themselves in America just like thousands of Jews had before them.

Solomon Tamkin was my father's grandfather. What he could not possibly have known when he arrived as a teenager was that he was entering at the end of an era. My great-grandfather came to this country toward the end of a half century that saw a dramatic increase in Jewish immigration to the United States. It was a half century that changed both the composition of American Jewry and other Americans' reactions and responses to it.

The first Jews to come to the United States, back in the country's earliest days, were majority Sephardic, originally from the Iberian Peninsula but from there traveled to southeastern Europe, the Levant, and North Africa, among other places. (It should be noted, though, that some consider Sephardim based on not ethnolinguistic categories but an approach to Torah informed by Spanish scholars and the Spanish liturgy.) Sephardic settlers came to New Amsterdam from Brazil in 1654. But Sephardim were shortly thereafter joined by Ashkenazic Jews, who originated in the Germanic lands; many then moved to eastern Europe and what was then the Russian Empire. By 1730, there were more Ashkenazic than Sephardic Jews in what would be the United States, though early synagogues still followed Sephardic rituals and customs.[2] Jewish congregations older than this country are still alive and at work today: In the late summer of 2021, while visiting Savannah, Georgia, I stopped by the synagogue of Congregation Mickve Israel. The congregation dates back to 1733, and its first synagogue was built in 1820. When I was there, I was greeted by a sign on the door instructing bat mitzvah guests to use a side entrance. Conversely, a Reform synagogue in Charleston, South Carolina, first built for a congregation in 1840, now has in front of it a plaque that acknowledges that its construction was overseen by a Jewish builder who used slave labor.

Under the Naturalization Act of 1790, only free white persons could become citizens of the United States.[3] Rights and privileges—including the privilege to have one's political place in the country fully recognized—were tied to whiteness. In the beginning of this country's history, Jews were a relatively small number of people who lived among white, Christian America—who could own enslaved people, and were not themselves at risk of being enslaved, to put it in the pre–Civil War context—and just happened to worship differently. In fact, Judah P. Benjamin, who was born to Sephardic Jewish parents in Saint Croix and raised in South Carolina, became a rich slave owner, a senator from Louisiana, and eventually the Confederacy's attorney general and secretary of state, making him the first Jewish person to rise to a cabinet position in North America.[4] The first Jewish person to hold a cabinet position on this continent, in other words, had his position among those who wanted to keep slavery, not among those trying to end it. (It should be noted, though, given the persistent myth that Jews controlled the slave trade, that Jews, who made up a small portion of the population, also made up a small portion of slave owners: according to historian David Brion Davis, in the American South in 1830, of the 45,000 slaveholders who owned 20 or more enslaved people, 120 were Jewish; of the 12,000 who owned 50 or more enslaved people, 20 were Jewish.)[5]

But the population did not stay sufficiently small or similar in character for this perception of Jews in the United States to persist. In 1820, there were some 4,000 Jewish people in the United States. In 1880, there were 250,000, a quarter of whom were in New York City.[6]

Beginning in the 1860s, early industrialization and urbanization forced Jews from their traditional positions in eastern Europe. Jobs like traders and peddlers and artisans disappeared, and there weren't enough factory positions to make up for jobs lost. What's more, after 1881, violence against Jews was both systematic and encouraged by governments, particularly in Russia.[7] The empire had long taken an exclusionary, oppressive attitude toward Jews, but it was then, following the assassination of Tsar Alexander II, that pogroms began.

My great-grandfather Herman's father—my father's mother's grandfather—was killed in a pogrom. His mother then took the children to the United States. It was discovered on their arrival that she had tuberculosis, and so while the children went to live with a relative, she was sent back. Herman, years later, went back to try to find her. She and everyone else he knew in his youth were gone.[8]

Herman's mother was one of thousands killed in pogroms. There was the Kishinev pogrom in 1903[9] and a subsequent wave of pogroms that lasted until 1906, the year that Jewish leaders admitted there had been too many pogroms to count. Eight hundred Jews died in Odessa alone. Violence against Jews tended to be particularly bad after Russia experienced a defeat, as was the case in 1906 after the Russo-Japanese War.[10]

Due to economic degradation and fear for their own lives, roughly a third of all eastern European Jews who came to the United States came between 1880 and 1924. Between 1881 and 1892, some 20,000 arrived each year. Between 1892 and 1903, that number shot up to 37,000. And 76,000 arrived between 1903 and World War I.[11]

This changed the composition of American Jewry—not only were there now far more Ashkenazim than Sephardim, but those Ashkenazim were largely from eastern Europe, as opposed to the mid-nineteenth-century immigrants who had come from Germany and central Europe. The latter group were wealthier and eager to acculturate quickly, to worship privately, and otherwise blend in with their new countrymen. And they set up groups to try to help the new Jewish immigrants do the same. For example, in 1893, Jewish philanthropists set up Educational Alliance on the Lower East Side of

Manhattan. The alliance worked to provide religious education, skills with which people could get jobs, preparation for civil service exams, and English classes.[12] The story of older, wealthier immigrants feeling shame over their counterparts who came over later with less money, less status, and perhaps less eagerness to immediately assimilate is hardly unique to Jews, but it is indeed part of the American Jewish story, too. Already, there was a push and pull over who got to dictate the right way to be Jewish in America.

Part of the problem for more acculturated Jews was that, as more Jews from more markedly different cultures came to America, Americans generally would have more difficulty seeing the Jewish population as these more established Jews wanted to be perceived: Americans who just happened to have a different religion. But it should be said that, even if Jews were simply to present as people who happened to pray differently, Jews in America were still not a monolith.

Some Jews, faced with this overwhelmingly non-Jewish environment, converted. For example, in the obituary of Irving Rosenthal, born in 1863 in New York, the stated reason for his conversion to Catholicism was that he was beaten up by Irish Catholics in his youth and so thought that it was a religion to which it might be worth converting. He took the last name Bacon, after his favorite doctor of the church. He even spoke with a Vatican representative on their visit to the United States, though this was less because of his religion and more because he spoke fluent Latin. (Granted, Irving is an atypical example of a convert, as, before his death, he converted back to Judaism, meaning that he died that rarest of things: a Jewish Bacon.)[13]

Some Jews felt that the conditions were not right in America for one to be a Good Jew. There were rabbis in eastern Europe who actually warned people not to go to America out of fear that being there would cause them to lose their faith.[14] One man I interviewed for this book, a forty-nine-year-old named Max Winter, shared that his great-grandmother had been the only one in her family to leave what she would only refer to as "the old country," so afraid were her family members that they would not be able to keep kosher in America. They were killed in the pogroms.[15]

And some also felt that they needed more structure and rigor and rules in order to worship properly. A number of Orthodox congregations tried to achieve this by bringing in an authority figure. In 1888, some decided to bring over Rabbi Jacob Joseph from Lithuania. He was supposed to be the chief rabbi for New York's Jews, a sort of Grand Rabbi, leading a rabbinical court and overseeing Jewish education. He was also supposed to take charge of kosher supervision. This project, however, did not go as intended; even Orthodox Jews in America, who ostensibly would have been the most open to greater structure imposed on worship, refused to accept that this imported rabbi was suddenly in charge. The Jews of New York, even back in the late nineteenth century, were too diverse and spirited to universally accept the rabbi. The congregations stopped paying him after several years, proposing that the butchers he was meant to supervise do so instead. Rival rabbis emerged. Joseph died defeated, though thousands came to his funeral in New York in 1902 (they were harassed, and the police officers who arrived sided with the attackers, not the mourners).[16]

This was a more significant moment for American Jews than those who rejected Rabbi Jacob Joseph could have known, and arguably changed the entire course of American Jewish history. There are still, of course, preeminent rabbis in particular communities in the United States, but there is no one person who can come anywhere close to saying with any legitimacy that they speak for America's—or even New York's—Jews.

But rejection of one figure and one source of authority didn't mean that Jews were rejecting Jewish religious structures. In places with high concentrations of Jews—namely, in New York—some tried to create a sense of Jewishness even outside of synagogues. Kosher butcher shops and bakeries popped up in Jewish neighborhoods. Communities constructed *mikvaot*, or Jewish ritual baths, and opened schools modeled on *cheder*, which taught Hebrew and religious knowledge.[17]

Still others evolved within the religion, or rather forced it to evolve for them. Many Orthodox Jews were firmly against adapting to the United States or to more modern times. The rabbis of the Agudath ha-Rabbanim, or Union of Orthodox Rabbis, founded in 1902, argued against English-language sermons and in favor of Yiddish-language instruction in Jewish schools.[18] But elsewhere, things were "Americanizing" rapidly. For example, the Jewish Theological Seminary in New York City was founded in 1886 as an Orthodox institution, but over the 1910s and '20s it became a bastion of Conservative Judaism, which stresses that Jewish people should both study tradition and history and engage with the here and now.[19]

Arguably no sect, however, put as strong an emphasis on the evolution of the religion as the Reform movement. Reform Judaism began in nineteenth-century Germany and the general principle, as the name suggests, is that Judaism is constantly changing and shaping and being shaped to fit the lives of its adherents. A major moment for the Reform movement came not in Germany but in the United States, and specifically in Pittsburgh.

The Pittsburgh Platform of 1885 called for Jews to accept a modern approach toward their faith. The platform sought to emphasize ethics, not ritual, and included a recognition of God but also said, "We hold that all such Mosaic and rabbinical laws as regulate diet, priestly purity, and dress originated in ages and under the influence of ideas entirely foreign to our present mental and spiritual state. They fail to impress the modern Jew with a spirit of priestly holiness; their observance in our days is apt rather to obstruct than to further modern spiritual elevation."

The Pittsburgh Platform's authors deemed it "our duty to partici-pate in the great task of modern times, to solve, on the basis of justice and righteousness, the problems presented by the contrasts and evils of the present organization of society."

The Pittsburgh Platform also explicitly said that Jews were not a nation but a religious community, "and therefore expect neither a re-turn to Palestine, nor a sacrificial worship under the sons of Aaron, nor the restoration of any of the laws concerning the Jewish state."[20] This speaks to the lack of following that Zionism—the movement to establish a Jewish homeland in Palestine—had in late nineteenth-century America, but so, too, does it speak to the reality that Jews in the United States at this time conceived of themselves as Americans of a different faith, and that was all.

But the Pittsburgh Platform was a turning point before a turning point. It was penned in 1885, at the very beginning of the increase in Jewish immigration. Its authors could not have imagined how the next forty years would change the way in which the country per-ceived Jews, and how increased immigration would challenge the conception of Jews and Jewishness they had just outlined.

In chapter two, we will examine the reasons that Jews in the United States were ambivalent toward whiteness culturally, but what must be understood in the context of immigration is that Jews from Europe or of European descent in the United States who were in positions of leadership or power were not ambivalent about being considered white legally. Whiteness meant that one could come to America, claim a spot, put down roots. To be white meant that one could become a citizen; to be anything else threw that, and one's legal place in the United States, into doubt.

Jewish leaders knew this especially well, and so were cognizant of downplaying Jewish distinctiveness when speaking to or with non-Jews. Well-known attorney Simon Wolf generally referred to

"American citizens of Jewish faith," though in Jewish publications he sometimes referred to the Jewish race.[21] If Jews wanted to grapple with whiteness or think of themselves as a people privately, fine, but what Wolf did not want was for Jews to be considered special or different or apart from the white, Christian, American mainstream at large. Race is mutable, and race is a construct, but here, in the United States, race was the difference between having the right to immigrate and become a citizen and claim space and not having any of that certainty or status.

More established and acculturated Jews tried to convince the Jewish masses—many of whom did see themselves as distinct—that they needed to present themselves to the non-Jewish public as more white than ethnic, more all-American than other, dissuading them from forming "Hebrew" political and workingmen's clubs.[22]

Even those who felt Jews to be distinctive were, in some cases, wary of straying too far from whiteness. While Zionists were typically among the first to point out Jewish racial distinction, Louis Brandeis, who led the US Zionist movement from 1914 to 1921, and who then became the first Jewish Supreme Court justice, said Jews were held together as much by "conscious community of sentiments, common experiences, and common qualities" as by race. Max Margolis, a Hebrew Union College professor in the Zionist movement, pointed out that while there were particularities in Jewish hair, for example, "if the color of skin be had in view, we belong to the whites."[23]

Of course, stressing these particularities as Jewish conveniently overlooked that Jews whose families come from Portugal—or Greece, or Turkey, or the Ottoman Empire—may have different hair than those from central and eastern Europe. The argument about Jews and whiteness being put forth assumed that the paradigmatic Jews were Ashkenazic Jews.

Not all Jews who were not Ashkenazic *wanted* to walk the same path as Ashkenazic Jews. Syrian Jews, for example, who also came over in the early 1900s, actively maintained an insular community.[24]

Even though Sephardic Jews were the first Jews in America, some found that their very Jewishness was questioned by many of their new Ashkenazic neighbors. They looked different; their prayers were pronounced differently; they didn't speak Yiddish.

Aviva Ben-Ur's *Sephardic Jews in America: A Diasporic History* opens with the story of Albert Amateau, who came to New York from Milas, a city on the Aegean. When he arrived in the United States at the age of twenty-one and sought out boardinghouse proprietors, Ashkenazic landlords turned him away. One even demanded to see whether Amateau was circumcised, and, when he showed that he was, dismissed the young man as a probable Muslim.[25] He had to find a hostel run by Sephardic Jews.[26] In the pre–World War I period, Ashkenazic Jews on the Lower East Side petitioned Mayor William Jay Gaynor to "remove these Turks in our midst."[27]

Just as American ideas of whiteness and race and who got to lay full claim to being American were internalized by Ashkenazic Jews, the idea that Ashkenazic Jews were the real Jews was internalized by some Sephardic Jews. In early-twentieth-century histories, Sephardic Jews referred to themselves as "Sepharadim" and Ashkenazic Jews as "Jews."[28]

This distinction was also reflected in treatment of non-Ashkenazic Jews by the state. When, in 1899, the Bureau of Immigration started recording Jews as "Hebrew," many Jewish immigrants from the Levant were not being processed as Jews. And in 1910, when the US census started including mother tongue for "foreign white stock," it used "Yiddish and Hebrew" for Jews, but not Ladino, or what was once known as Judeo-Spanish, or other non-Ashkenazic Jewish languages.[29]

One might imagine that Ashkenazic Jews, in this period of thinking through what Jewishness was, could have expanded the definition to include non–central and eastern European Jews. Evidently, many chose not to. Perhaps this was because they felt that doing so would have threatened them. Rights and privileges and one's status as an American citizen, after all, were tied to whiteness. And so Jews felt they needed to be able to prove theirs. This was a process that would have been complicated by allowing that Sephardic Jews were *really* Jewish. Sephardic—and Mizrahi—Jews from the Muslim world would have been a liability. Perhaps by introducing other, less acceptably white Jews to white, Christian America, all Jews might be considered inferior.

This would have been a reason. But it would not have been an excuse. I thought, learning of these stories, about my great-grandparents. I thought about how hard it must have been for them to make a life for themselves in this new country. I thought about how they crammed together in apartments, and somehow made their way, and fought to be able to provide a life for their children. I thought of how important their Jewishness was to them throughout that process. They kept their Yiddish, their food (even if it was boiled chicken and schnapps), their prayers. I wondered how much harder life would have been if all of that—their identity—had been taken away at the hands of other Jews. And I thought about how Ashkenazic Jews grappled over their self-identification, and whether they were a race or a religion or an ethnicity or a people, and how they decided how they wanted to represent themselves to America, and how they then took that from other Jews. If America failed to understand and so attempted to flatten Jewish identity, then, it must be admitted, some Ashkenazic Jews failed and flatted other Jewish identities, too.

The American Jewish Committee was established in 1906 to combat antisemitic attitudes and behaviors. It was not the only group in this time period to do this—the Anti-Defamation League was started in 1913 for similar reasons[30]—but the members of the AJC were considered, and indeed considered themselves, the Jewish elite (or, as historian Jonathan Sarna put it, "patrician-dominated").[31] And one of these elite Jews, Cyrus Adler, wrote to his colleague Mayer Sulzberger with a warning. Once Syrians and Japanese people were considered non-white, he said, "it will not be a very far step to declare the Jews Asiatic." The American Jewish Committee pressed both Congress and the Census Bureau to remove European immigrant "races," including Jews, from the census, and in 1910 Jewish leaders picked up the fight to ensure the continued eligibility of Jews for naturalization.

A sympathetic Jewish congressman from New York, Henry Goldfogle, pushed a bill through the House that proposed "Asiatics who are Armenians, Syrians, or Jews" not be barred from naturalization. Louis Marshall with the American Jewish Committee was appalled. Jews weren't supposed to be Asian but nevertheless eligible for naturalization; they were supposed to be white. Marshall scuttled the bill, which never made it to the Senate floor.[32]

On the other end of the spectrum, there was the reaction to Simon Wolf. In 1909, Wolf, a Washington attorney representing the Union of American Hebrew Congregations, testified before Congress's US Immigration Commission that "Jewish" should not be a racial classification. He denied all racial distinctions. But as a result, he seemed inconsistent—Wolf tripped up on a line of questioning about Benjamin Disraeli, the late British prime minister who was born Jewish and then was baptized. Wolf said that, religiously, he ceased to be Jewish. Senator Henry Cabot Lodge seized on the fact that if someone ceased to be Jewish *religiously*, then there was necessarily another way in which they could continue to be Jewish. The committee left unimpressed with, and unconvinced by, Wolf.[33] The tide was already turning. Jews were seen as not quite white, not quite wholly American, not quite able to assimilate.

Part of the idea of Jewish separateness was that one of the ways in which American Jews at the time retained a distinctively Jewish identity was through their insistence that Jews only marry other Jews. Between 1908 and 1912 in New York City, only 1.17 percent of Jews intermarried. Only Black Americans in New York City had a lower rate of intermarriage.[34] (Syrian Jews took a particularly strong stance against intermarriage; in the 1930s, they adopted an edict saying that converts could not be considered Jews.)[35]

In 1908, British Jewish playwright Israel Zangwill came out with his play, *The Melting Pot*, which was about the union of a Jewish and a non-Jewish immigrant, both of whom came to America from Kishinev. The play was popular. Theodore Roosevelt, for one, loved it. Some American Jewish leaders felt the need to respond to the suggestion that, as Eric Goldstein, author of *The Price of Whiteness*, puts it, "racial amalgamation was a prerequisite to becoming true Americans."[36] For some, that meant that they came out and said that intermarriage was fine. As a character in one of the stories by Anzia Yezierska, herself a Jewish immigrant to the United States, wrote, "America is a lover's land."[37] To marry for love, not religion, was all-American.

But others tried to come up with justifications for denying the validity of intermarriage. Rabbi Mendel Silber argued that Jewish women were meant to be "priestesses of home," and they simply couldn't be replaced by non-Jewish women. Even those in the ostensibly forward-looking Reform movement were against intermarriage, trying to argue that opposition to intermarriage was based only on religious, and not racial, grounds. Still, the 1909 Central Conference of American Rabbis could not manage to muster support to pass a resolution prohibiting rabbis from performing intermarriages. Instead, they came up with a statement that said that, according to Jewish tradition, such unions were looked on more negatively.[38]

That negativity was felt by those who entered such unions. Mary Antin, author of the autobiographical Jewish immigrant memoir *The Promised Land*, married a German Lutheran. Antin wrote to Zangwill himself, "I . . . hope that none of my old friends will think that I can spare them now. I want them all as much as ever, particularly since I have lost many to whom my marriage was displeasing on religious grounds. They might find these reasons unfounded if they could realize that I have not changed my faith."[39]

This resistance to intermarriage, coupled with race science and xenophobia, would be taken by some Americans as a refusal to assimilate, a perception that would contribute to a very real threat to the legal status of Jews in the United States.

In 1914, the empires of Europe went to war. In 1917, the United States joined in, fighting on the side of its French and British allies. For young Jewish men, as for many young people of newer immigrant groups, the war was an opportunity. Surely, they thought, by serving their country, they could become more quickly accepted into the American mainstream.

"The ideal for which we are fighting is to suppress the great bully and outlaw among the nations—the German government," said Mordecai Kaplan, the father of Reconstructionist Judaism and rabbinical scholar, adding, "as Jews . . . we owe it to America to stand by her in her hour of trial."[40]

But America would soon decide it did not want to stand by the Jews.

In 1918, Kaplan wrote in his diary that though the Great War had come to an end, "Now we shall have a war of classes." He added, "Social revolution is in the air these days."[41]

Jews (or at least many Jews) had already been worried about being seen as insufficiently American. Now the air of social revolution brought a whole new subset of problems. Jews, who were perceived as being overrepresented in the ranks of Socialists and Communists, were worried that they would also be cast as subversive. They had good reason to worry. Attorney General A. Mitchell Palmer wrote in a letter to the aforementioned Simon Wolf, "It is impossible to state whether or not these persons still adhere to the Jewish faith, for previous investigations by the [Justice] Department have shown that many . . . members of the Jewish faith, have renounced that faith and are at the present time in no way connected with any religion."[42] Perception of Jews was changing. Some in power now saw them not as a set of people who adhered to certain religious beliefs but rather as a group of potential radicals and revolutionaries, which in turn attracted negative government attention.

At the same time, Jews' own relationship to worship was changing. Jewish observance was in a transitional period. According to the 1926 US Census of Religious Bodies, there was one synagogue for every 1,309 Jews (compared to one church for every 220 Christians) and the average Jewish child received Jewish education for just two years.[43] There was new music, new style, and, as Jerome Mintz wrote in *Hasidic People*, "acceptable ways for men to pass time in the evening and on [Shabbos] other than at Torah study."[44] While the subsequent decade—from 1925 to 1935—was considered a time of "religious depression"[45] in the United States generally, and not only for Jews, statistics from the time suggest it was more pronounced for Jews than it was for Christians.

This isn't to say that Jews felt themselves as less Jewish. There was still an intensely "Jewish" feel to life in at least some cities. "The dominating characteristic of the streets on which I grew was Jewishness in all its rich variety," Vivian Gornick wrote. "We did not have to be 'observing' Jews to know that we were Jews." But Jews as a group or people were less comprehensible to other Americans than Jews as religious practitioners.[46]

This version of Jewish life came out of a very particular time and place. The late nineteenth century saw the emergence of the General Jewish Labor Bund in Lithuania, Poland, and Russia. The Bund was a

secular Jewish Socialist party, a response to the antisemitism that was
so tied up with the capitalist system within the Russian Empire and
to the repressive political conditions of the day.[47] Bundists and those
sympathetic to their cause who immigrated to the United States
came to a very different political system and situation but brought
those values with them. "Yiddishkeit," a term that attempts to cap-
ture the notion of Jewish life or way of being Jewish, thus took on a
Socialist streak for many in the United States.

This was a political project, but it also had a cultural and linguistic
component. That streak appeared in almost every element of Jewish
life. There was the language, for one. But Yiddish also brought with
it a culture and way of life. There was the Yiddish press, which, at its
height, was a force in American media. The most famous publication
was *Der Forverts*, or *Forward*, which was founded in 1897 by Abraham
Cahan, an immigrant from Russia. In 1915, it sold 150,000 copies a
day in New York.[48] By the 1920s, circulation of the Yiddish *Forward*
was greater than that of the *New York Times*.[49]

And there were other organizations, too, ones around which left-
leaning Jews could orient their lives. The year 1918 saw the establishment of the Sholem Aleichem Folk Schools, named for arguably
the most famous Yiddish-language writer ever to live, which were
Yiddish-language schools that taught Jewish culture over religion.
In 1900, the Socialist Arbeter Ring, or Workmen's Circle (today, the
Workers Circle), was established to help Jewish immigrants, providing them with life insurance and unemployment relief, but also with
dances and educational classes. "Most of them had rejected religious
observance, but they continued to embrace Jewish traditions, holidays, and texts to guide their activism," Ann Toback, CEO of the
Workers Circle today, told me. Jewishness was a part of the group—
their own kind of Jewishness.[50]

In other words, in the early twentieth century, there existed in parts
of the United States with heavily concentrated Jewish populations

whole Jewish worlds that revolved around culture, not religion. That might have brought on problems for Jews even if everything else had stayed the same. But everything else didn't, and the anxiety caused by a rapidly changing world—one with industrialization and immigration and tensions in Europe—meant dominant groups frequently targeted minority groups to feel in control of a changing American landscape. Non-Jewish Americans often lashed out at American Jews and Jewish immigrants and used "science" to justify doing so.

In his 1911 work *Heredity in Relation to Eugenics*, Charles Davenport wrote that Jews—and specifically Jews from eastern Europe—were marked by "intensive individualism and the ideals of gain at the cost of any interest." Jewish and Italian "blood" in the United States, he wrote, would make Americans "darker in pigmentation, smaller in stature, more mercurial . . . more given to crimes of larceny, kidnapping, assault, murder, rape, and sex-immorality." These ideas were not unique to Davenport; they were widely disseminated and popular at the time.[51]

Obviously, politicians serving their good white American constituents couldn't have intensive individualism or the potential for smallness of stature arriving at Ellis Island. And so American lawmakers, like many of their constituents, became fond of the idea that some European immigrants—those from northern and western Europe—were preferable to others—namely, those from southern and eastern Europe.

Prior to this point, private institutions often discriminated against Jews. But that discrimination, as Hasia Diner writes in *The Jews of the United States*, generally speaking, "stopped at the gates of government": the US government treated Jews as white, even if private institutions did not.[52] This does not mean there was no discrimination at the government level against Jews. There was a stipulation in the state of Maryland's constitution that prevented Jews from holding public office until 1826 (at which point the state legislature passed a law, known as the "Jew bill," that annulled the state's constitutional requirement that state officers be Christian).[53] There was General Ulysses S. Grant's 1862 issuance of Order 11, which said that Jews had to leave the "department of Tennessee." (Grant later became the first sitting US president to attend a Jewish service.)[54] But, broadly, under federal law, Jews in the United States had the rights and privileges that came with being a white American.

Yes, Harvard proposed a quota on how many Jewish students could attend, and yes, Harvard president A. Lawrence Lowell wrote, "The summer hotel that is ruined by admitting Jews meets its fate . . . because they drive away the Gentiles, and then after the Gentiles have left, they leave also,"[55] and yes, Jewish admittance was capped at other institutions of higher education and Jews were banned from their social clubs. A broadside spotted at New York University from 1923 read, "Strictly Kosher—Must not APPLY HERE. SCURVY KIKES ARE NOT WANTED. At New York University if they knew their place they would not be here. Make New York University a White Man's College."[56] When these quotas were in place in the interwar years, many Jewish Americans went instead to, for example, "Jewish Harvard"—that is, City College of New York—and founded their own country clubs and fraternities and sororities and resorts, where, per the *Forward*, they "had their revenge on the Gentiles who didn't want to accept them." All of this served to reinforce the segregation of Jews from the rest of American society.[57] "Even as they gloried in being part of the larger American culture they remained firmly rooted in a subculture that consisted largely of Jews," Jonathan Sarna wrote in *American Judaism: A History*.[58] But all this was unequal treatment at and by private institutions, not by the United States government.

There were also a number of incidents where violent bigotry unquestionably played a role. In 1913, when thirteen-year-old Mary Phagan's body was found by a night watchman in Georgia, police blamed Leo Frank, a Jewish man. Frank was sentenced to death. Adolph Ochs, owner of the *New York Times*, tried to help Frank with favorable coverage, but, perhaps unsurprisingly, a Jewish media magnate trying to improve coverage for another Jew backfired where public sentiment was concerned. The governor reduced Frank's sentence to life imprisonment. Then, in 1915, a mob of twenty-five men overpowered the prison guards, drove Frank a hundred-plus miles away, and lynched him.[59]

Though this is a gruesome example of a gross miscarriage of justice, it is not regular, systematic discrimination by the federal government. There was, broadly speaking, cultural discrimination against Jews, and that was at times reinforced and reflected by institutions like the justice system. But that is very different from Jews in America being legally classified and treated differently. Discrimination against Jewish people was not baked into the country's founding documents as, say, slavery was. There was prejudice, and there was hatred, and there was violence, but it was not because of the law as it was written.

The *Dearborn Independent*, published by Henry Ford, father of the mass-produced automobile, was also not a government entity. In his publication, Ford ran a series of articles on the various ills committed by the "International Jew," whom he faulted for manipulating the American financial system and corrupting Anglo-Saxon values with Hollywood and music. He also called Black American jazz "Yiddish moron music"[60] and drew heavily from *The Protocols of the Elders of Zion*, which was originally published in the Russian Empire and falsely claimed there existed an international Jewish conspiracy that wanted to control the world by manipulating the media, banking industry, and economy. (Incidentally, Henry Ford sought out the friendship of Detroit rabbi Leo Franklin, offering a neat early-twentieth-century example of political actors trying to keep Jewish people—just not too many—close to justify their own sentiments.[61] History before and since has been littered with people who say that they cannot possibly be prejudiced against Jewish people because they have Jewish friends. US vice president Mike Pence, in 2020, said that Trump couldn't be a white supremacist, pointing to the president's Jewish grandchildren. And when he did, I thought of Henry Ford and Leo Franklin.) But Henry Ford was not an elected official.

Then, in 1919, there was a shift. Proponents of cultural and individual discrimination discovered an avenue to transform their prejudice into law. The Ku Klux Klan, back to the foreground of American political life with a vengeance and a newfound passion for antisemitism, found an ally in Albert Johnson, the new chair of the House Immigration and Naturalization Committee.[62] Five years later, in 1924, twelve years after my great-grandfather Solomon came to Massachusetts, Americans would decide that they had had their fill of Solomons and Fannies, that it was time to turn off the tap of Jewish immigration to the United States. The decision would shape the face of American Jewry, Judaism, and Jewishness for the next hundred years.

Though there were not as many Ashkenazic Jews coming to the United States as there were in the prewar period, fighting among Russian, Ukrainian, and Polish forces and a series of pogroms did see 119,000 immigrants from central and eastern Europe arrive in America. Lawmakers demanded immigration be suspended. Johnson also pointed to a State Department document that said that "abnormally twisted" and "unassailable Jews," "filthy, un-American, and often dangerous in their habits," were coming to America.[63]

As Jia Lynn Yang outlines in *One Mighty and Irresistible Tide: The Epic Struggle over American Immigration, 1924–1965*, Johnson and company didn't want immigration from southern and eastern Europe cut down a little; they wanted it cut down a lot. Johnson wanted quotas based on the composition of the United States according to the 1890 census, which would heavily favor northern and western Europeans. "The use of the 1890 census is not discriminatory. It is used in an effort to preserve, as nearly as possible, the racial status quo in the United States. It is hoped to guarantee, as best we can at this late date, racial homogeneity in the United States," Johnson wrote.[64]

Jewish leaders and lawmakers railed against the proposed law. A thirty-five-year-old New York lawmaker, Emanuel "Mannie" Celler, spoke at a Brooklyn Jewish Center, pointing out that English immigrants took longer on average to apply for citizenship and said that Labor Secretary James J. Davis, a Welsh immigrant who had the gall to become an immigration opponent, was promoting anti-immigrant propaganda.[65] William Edlin, the editor of *Yiddish Day*, testified before the House Immigration and Naturalization Committee in 1923 that "every worthy man, woman, or child is entitled to be admitted into this country," although he then said he was speaking only on behalf of "Caucasian race" and "The Chinese, Hindus, and other races do not have those things that we call civilization, and I look upon those people as too far from us for assimilation purposes. . . . We are pleading only for the white population and for no other."[66] Similarly, Czech-born Jewish Democrat Adolph Sabath, who served on the House Immigration and Naturalization Committee, said the bill would be the "first instance in our modern legislation for writing into our laws the hateful doctrine of inequality between the various parts of our population," a quote that makes it seem Sabath either did not care about or had somehow been unaware of the Chinese Exclusion Act passed some forty years prior and also of the US constitution.[67]

Still, though American Jews may have been treated as not white in their day-to-day experience, they were considered white officially, which is to say, by the US government. Solomon and Fannie and my father's mother's parents, Sally and Herman, were all recorded as "white" on the 1920 census.[68] By this all-American logic, they should not have been discriminated against. But, as the US government was about to make very clear, there was whiteness and there was whiteness.

There was, of course, high-profile pushback. Israel Zangwill, the British author of the play *The Melting Pot*, opined, "If you create enough fuss against this Nordic nonsense, you will defeat this legislation. You must fight against this bill; tell them they are destroying American ideals. Most fortifications are of cardboard, and if you press against them they give way." This had the opposite of the intended effect. Johnson's committee held this up as proof of outside interference with US legislation.

"You are declaring the incapacity of America to Americanize," Rabbi Stephen S. Wise, a giant of Reform Judaism, pleaded with the committee.[69]

Wise had misunderstood; the legislators didn't *want* an America that could proudly Americanize. They wanted an America for those *they* believed belonged in America. The law, Pennsylvania senator David A. Reed wrote in the *New York Times* that April, "will mean a more homogeneous nation, more self-reliant, more independent and more closely knit by common purpose and common ideas."[70] Like President Donald Trump almost a century later, the legislators of 1924 wanted more immigrants from places like Norway.[71]

There was support for the bill from elsewhere in government, too. John Trevor, head of the New York branch of the US Army's Military Intelligence Division, decided that the way to be rid of Communist subversion was to decrease the number of Jewish people entering the country. He came up with a clever, if cynical, way to respond to the criticisms of Jewish and Italian groups. Basing quotas on the foreign-born population left out the people whose ancestors had shaped America's founding, and so he proposed that the government base the quotas instead on the ancestry of the country (that is, the part of the country that was of European descent) in 1920. Trevor's calculations said that this meant that 84 percent of immigration would be reserved for immigrants from northern and western Europe. Senator Reed agreed.[72] The law passed, and President Calvin

Coolidge, who had pledged that "America must be kept American," signed the Johnson-Reed Act into law on May 26, 1924.[73]

The tap of Jewish immigration to America was turned off. Only a trickle would remain. Even Rabbi Wise could not have predicted the consequences.

The Johnson-Reed Act hastened assimilation. With hardly any new immigrants, American Jews came to resemble more closely other Americans very quickly. Yiddish language and culture went into decline, though its food-based traditions remained. American Jews began moving out of the compact, "ethnic" urban neighborhoods into other parts of the cities and their environs. Jews in New York moved from the Lower East Side to Brooklyn and the Bronx, while Jews in Boston moved from the city's North End and Beacon Hill to Roxbury and Dorchester. Even in these new neighborhoods, though, many lived lives that were apart from their non-Jewish neighbors.[74]

Dorchester is where Solomon lived when my father was growing up, and where my father would visit a "packed, messy, and frenetic" triple-decker apartment complex that included Fannie's mother, Bubbe Rosa, who ran the show (she was "always" with them when Solomon and Fannie were dating, he later lamented),[75] and Fannie's sister, Bessie, and where Solomon and company continued to speak Yiddish and eat, per my father, "all the big, sloppy East European food, boiled chicken, just horrendous food."

The house had grape vines on lattices in the front; it backed out into Franklin Field, where his son would witness interethnic brawls and his grandchildren, including my father, would one day play. Still, a move does not necessarily make an assimilated American: Solomon invited Jehovah's Witnesses who came calling to convert him into his apartment for schnapps, over which he would tell them about how much he loved being Jewish and that it was they, not the Jews, who were on the wrong track.[76]

My father's mother, on the other hand, grew up in an all-American house in Schenectady, New York, and though her family was more religious than Solomon's, they were less outwardly obviously Jewish. My nana's sister, my great-aunt Beverly, told me, "By the time I came along, we were very, very, very, very Americanized. Very much so. My father, who was also in the meat business, had done very, very well." She and my nana grew up in Schenectady and in 1934 her father bought "this beautiful home, which is still there and it's lovely" in "a very, very lovely neighborhood with a very good school system, one of the best in the state."

It was not, my great-aunt told me firmly, "in any Jewish neighborhood. Sometimes you'd get these little ghettos. There were all different people on the block. I grew up in a very lovely home." Her mother, my nana Sally, wore (according to my father's recollection) a blond wig. It may have partially been for religious purposes, he said, but he thinks that in part she just wanted to look her best for Schenectady society.[77]

And my nana and great-aunt's story was not uncommon; many Jewish families were moving out into lovely homes in suburbia around this time. They sometimes faced resistance when they tried to move into new neighborhoods. At first, some owners refused to rent to them. Some neighbors left their homes, not wanting to live next to Jews. But eventually the force was too great, and American Jews began building both suburban synagogues and the next chapter of American Jewish life.[78]

Jews also became still more entangled and a part of America's leisure activities—including America's favorite pastime. In 1933, Hank Greenberg, born to immigrant parents in New York in 1911, signed to play baseball with the Detroit Tigers. The next year, he sat out the game on Yom Kippur, the Jewish day of atonement, a move fans resented but also respected, as Edgar A. Guest captured in his poem on Greenberg's decision, published in 1934 in the *Detroit Free Press*.

The Irish didn't like it when they heard of Greenberg's fame
For they thought a good first baseman should possess an Irish name;
And the Murphys and Mulrooneys said they never dreamed they'd see
A Jewish boy from Bronxville out where Casey used to be.
In the early days of April not a Dugan tipped his hat
Or prayed to see a "double" when Hank Greenberg came to bat.
In July the Irish wondered where he'd ever learned to play.
"He makes me think of Casey!" Old Man Murphy dared to say;
And with fifty-seven doubles and a score of homers made
The respect they had for Greenberg was being openly displayed.
But upon the Jewish New Year when Hank Greenberg came to bat
And made two home runs off Pitcher Rhodes—
They cheered like mad for that.
Came Yom Kippur—holy feast day worldwide over to the Jew—
And Hank Greenberg to his teaching and the old tradition true
Spent the day among his people and he didn't come to play.
Said Murphy to Mulrooney, "We shall lose the game today!
We shall miss him on the infield and shall miss him at the bat,
But he's true to his religion—and I honor him for that!"[79]

Jews shaped America's music of the day and age, too. Benny Goodman, born to a poor family in Chicago in 1909, was enrolled as a child in a band at the Kehilath Jacob Synagogue. He was given a clarinet and went on to study music seriously. By the end of the 1920s, he was the King of Swing. But Goodman didn't just change the tunes; in 1936, he assembled a band in which Black and white musicians played together publicly for the first time.[80]

American Jews were not only becoming more involved in various forms of entertainment; they were shaping what entertainment in the United States looked like.

American Jews were also in the process of reshaping Judaism itself, changing it to fit their new lives in their new country.

Mordecai Kaplan—the rabbinical scholar who had worried in his diary about what was to come following the First World War—was particularly concerned with the question of how to reconstruct Judaism to make it workable for Americans. In 1934, he published *Judaism as a Civilization: Toward a Reconstruction of American-Jewish Life*, which argued that American Jews had split lives, divided between their existence as Americans and as Jews. He aimed to bring those two lives together, arguing that halakha (all Jewish laws and ordinances that have evolved since biblical times, taken together to regulate both observance and life) should have a "vote but not a veto" over how American Jews lived their lives. This was the emergence of Reconstructionist Judaism, a branch of Judaism that still exists today, that commits itself both to Jewish tradition and to finding meaning in contemporary life.

Not everyone was enthusiastic about Kaplan's project. In the 1940s, Kaplan published *The New Haggadah*, the book Jews use to guide through a Passover dinner. Kaplan's was a decidedly updated version. This Haggadah was popular, but came under criticism, for Kaplan had removed references to Jews as the "chosen people" and included an African American spiritual, "Go Down Moses." It was attacked by those more conservative than Kaplan. In fact, some Jews were not content to criticize, and in 1945 Orthodox rabbis gathered together in New York and burned copies of the book, turning Kaplan's Haggadah to ash.[81] (The inclusion and use of spirituals is also, today, debated in another context, as Black Jews have noted that non-Black Jews who sing the songs—particularly without fully appreciating or acknowledging their original context—are engaging in appropriation.)[82]

This incident is an almost perfect illustration of the meaninglessness of the term "Bad Jew," or rather of how its meaning is determined by whoever happens to throw out the term. Who, in this story, is the better Jew: the one who rewrote the Haggadah or the one who set it on fire?

Still, even with differences between various sects of Jews about how to practice Judaism in America, as Jonathan Sarna writes in *American Judaism,* "despite all the sound and fury of intra-religious conflict, many (though not all) Jews found themselves in agreement on a host of critical issues," including that "they sought simultaneously to be both American and Jewish."[83]

The Johnson-Reed Act, the piece of legislation that significantly restricted immigration from southern and eastern Europe, sped up acculturation as far fewer Jews were arriving, and American Jews were left both to assimilate and to grapple with what it meant to be a Jew in America.

But the Johnson-Reed Act corresponded with another phenomenon for American Jews, too: the rise of antisemitism in the United States.

The Johnson-Reed Act passed in 1924. The Great Depression came just a few years later, shaking America to its core in 1929.

Twenty percent of Jews in New York and 10 percent of Jews nationwide experienced losses because of the Depression, as did an even higher percentage of Jewish businesses and institutions. Beginning in 1930, every Jewish social services organization had more demand to meet with fewer resources. Jewish educational institutions were especially impacted: at one Connecticut yeshiva, students regularly went hungry, relying on donations from bakers and women who had chicken to spare; at one yeshiva in New York, instructors went unpaid for sixteen weeks, starving and suffering.

However, many Jews were self-employed and were not hit quite as hard as other Americans. Jewish organizations established agencies to handle the crisis: HIAS opened facilities to people who needed shelter. Synagogues opened their doors to the homeless. Additionally, American Jews assumed responsibility for helping each other and tried to find jobs for other American Jews. Perhaps for this reason,

Jewish unemployment in America's large cities was less than half the national average.[84]

Maybe the fact that Jews were not quite as badly off as others is the reason that antisemitism continued in the 1930s. Maybe it was that, in uncertain times, Americans were looking for a group to scapegoat, and it was easier to blame American Jews than to consider that they, too, were Americans impacted by an economic crisis. Whatever the reason, Jews were indeed thought by many to be responsible for the economic turmoil in the eyes of their fellow Americans.

In several Southern states, Jewish Communists were active in organizing Black workers during the Great Depression, and so were seen as instigators and troublemakers by local white Americans. In Miami in 1930, David Weinberg, a Jewish tailor and Communist, was singled out for "associating with Negroes."[85] (Meanwhile, many Black Americans in the 1930s blamed Jews for their suffering during the Great Depression, accusing, for example, Jewish store owners of exploitation and discriminatory hiring practice—as Goldstein writes in *The Price of Whiteness*, while this "overstated [Jewish] responsibility for the suffering of African Americans during the Depression years," the reality was that "many Jews had become complicit in the larger white power structure.")[86]

Mainstream white America proved eager not only to cast Jews as other but to cast that other as immoral and dangerous; not just as a revolting passive presence but as an affirmatively dangerous force for ill. Again, this wasn't new—Christians claimed Jews had brought about the Black Plague and slaughtered Jews before leaving for the Crusades in the Middle Ages—but rather an American manifestation of a centuries-old trope.

This myth was propagated by countless individuals, including Charles Lindbergh, who completed the first solo flight across the Atlantic in 1927, and who supported and spread conspiracy theories about Jewish responsibility for America's problems. As Fascist governments came to power in Europe, Lindbergh claimed that Jews would once again plunge the nation into war. The "greatest danger [Jews posed] to this country," Lindbergh said, was "in their large ownership and influence in our motion pictures, our press, our radio, and our Government."[87] Lindbergh, like Henry Ford, also had a few close Jewish friends, and said he was an admirer of the Jewish "race."[88]

An ocean away, in Europe, laws restricting what Jews could and could not do were passed by leaders and parliaments, who believed Jews were responsible for or benefitted from their losses in World War I. In the United States, antisemitism gained a larger audience via the airwaves when "radio priest" Father Charles Coughlin blamed Jews for the suffering of the American people.

Coughlin's radio broadcasts had an audience of over thirty million listeners. Coughlin argued in one November 1938 broadcast that Jews were responsible for Nazism and controlled US media. One month later, Rabbi Wise hit back against Coughlin's insincere query as to why Jews had not disavowed Communism by saying, "Coughlinism knows as well as you and I that not one hundredth of one percent of Jews outside the Soviet Union are communists. . . . No man, no movement, has the right to ask us Jews, to ask Jews as Jews, or any other social or religious group to make a vow or denial of any social, political, or economic theory of life. It is impossible. No one has the right to ask, no one has the right to answer and to speak for Jews."

"For the Jew," Wise concluded, "Coughlinism is a regrettable phenomenon. For the Catholic Church it is a disaster. But above all it is America's shame."[89]

Previously, arguing to Congress in favor of a more open immigration policy toward Jewish people, Wise had tried to appeal to America's better angels. But the angels had not answered. And so now he tried to shame them.

But in responding, he fell into Coughlin's trap. Though he said Jews could not be asked to disavow Communism, Wise, in a way, did, assuring listeners that the vast majority of Jews were not Communist. Even as he chided those in mainstream America, he tried to argue that Jews were indeed closer to them than to a radical fringe.

But Jews could not declare themselves moderate and therefore safe and gain acceptance from their fellow Americans. In fact, ironically, the more Jews moved into the mainstream and into American life, the more they were seen as having undue influence over society. There is one especially neat illustration of this no-win situation that is antisemitism: the New Deal.

Franklin Delano Roosevelt was, himself, a world away from the immigrant (or child of immigrants or even grandchild of immigrants) experience. Still, though he himself was not a Jewish immigrant or a descendant thereof, as governor of New York he made Herbert Lehman, son of one of the three Lehman Brothers, the Jewish banking family, his lieutenant governor. Roosevelt, who came up through New York politics, forged a close relationship between his party and American Jews,[90] some of whom were instrumental in crafting the New Deal, a series of programs and reforms intended to get America through the Great Depression.[91]

If this period saw American Jews step into their power as voters and even as policy makers, so, too, did it see another change: after World War I, even establishment Jewish leaders were referring to Jews as a "racial or ethnic" minority.[92] American Jews responded to rising antisemitism not only with fear but also with pride. While labor activist Bertha Wallerstein asked Jews in the magazine the *Nation* to fight for rights as "human beings without making a whole romance of the race,"[93] many others celebrated their status as Jews, pushing a pluralist vision of American society and stressing Jewish contributions to it.

In some cases, Jews used the same kind of faux science that had been weaponized against them to make the case of their distinctiveness. Dr. Nathan Krass, rabbi of New York's elite Reform synagogue, Temple Emanu-El, argued that intermarriage would spark a decline in "superiority of Jewish genius."[94] In 1933, Yiddish journalist and activist David Goldblatt wrote, "The Aryans may talk peace but they never show they mean it. War is in their blood." And Rabbi Wise, speaking before the National Association for the Advancement of Colored People, said, "I confess that I am a little prouder than I was a year ago that I am a Semite, prouder and happier than ever that there is not a drop of Aryan blood in my veins."[95]

Notably, this is the same Rabbi Wise who chided lawmakers, who said that America wasn't being left to Americanize. It wasn't that Wise was now saying that he wasn't American, but he was insisting that Americanness itself did not—could not—take Jewish identity away.

In some cases, pride and certainty in Jewishness just meant that Jews felt more comfortable expressing that they were Jews and standing up to antisemitism. They were being promised integration into mainstream America, or at least into its liberal variant. Some felt empowered to say that antisemitism was antithetical to America's values, which is to say that they felt capable of taking up enough space to assert their ideas about American values.

This was not lost on the most famous American politician of the era. FDR courted Jewish favor, and he did so effectively through inclusive language and also liberal policies that appealed to Jewish people. This led many Jews to move from the Socialist fringes and into the Democratic-voting mainstream; they did so because they liked FDR and what he was offering them, and because, in some ways, the liberal values of their Socialist parents appeared, at least to an extent, in FDR's platform. And FDR hired many Jewish people, bringing them in to staff his administration in greater numbers than before. They might not be able to get work at prestigious law firms because of the pernicious influence of anti-Jewish prejudice in white-collar industries, but they could work for the US president.

FDR even had a Jewish cabinet secretary: Henry Morgenthau, Treasury secretary.

Hiring Morgenthau went down badly with antisemites, a growing club that had no shortage of members in elected office. The New Deal was derisively called the "Jew Deal."[96] Martin R. Dies of Texas, chair of the House Special Committee on Un-American Activities, frequently accused Jews of being Communist sympathizers; he was especially angry with those Jews who supported civil rights. Senator Theodore Bilbo and Congressman John Rankin of Mississippi blamed "New York Jews" for the New Deal and slammed those Jewish members of Congress who were supportive of desegregation and labor reform.[97]

Despite their representation in the cabinet and White House, American Jews did not, in fact, have real political influence with Roosevelt, at least where Jews were specifically and directly concerned. One of the darkest tragedies of human history was about to demonstrate that.

When Arlette was four, her family left Europe. She'd been born in 1936 in Paris, France, the daughter of Jewish immigrants who had met in Belgium, three years after Adolf Hitler became the German chancellor. Her father's optimism remained in place during the beginning of Hitler's rule. The Germans would never invade France, he would tell her mother, and, besides, France had the Maginot Line.

But Arlette's mother did not share his optimism. Due to either great cowardice or great courage, as Arlette would later say, her mother insisted they leave. *We're not safe here. We have to get out of here.*

The family left for Portugal. There were quite a few other Jewish families there at the time; though Portugal's government was Fascist, this particular crop of Fascists was not especially interested in antisemitism. But Arlette's mother wasn't satisfied. *We're not safe here. We have to get farther away from them.*

The family took a boat; Arlette loved running up and down the deck, hiding from her parents. They were supposed to go to Canada, which was better for refugees, but stopped in New York in 1941. Her father liked the city, and because he made tools at the time, he was able to get the family visitors' visas. They moved to the Bronx, where Arlette was out of place with her white fur coat and muff. She told her mother she would never speak to her in French again; she was American now.

Arlette Sanders was my high school mentor. While interim head of the English Department, she'd take time after the school day to read and discuss books with me. One of them was *The Plot Against America* by Philip Roth, which imagines what would have happened to America's Jews had Charles Lindbergh beaten Franklin Delano Roosevelt in the presidential election of 1940.[98]

Arlette's family was lucky. By the most generous estimate, between 1933 and 1944, the United States admitted only 250,000 Jewish refugees.[99]

When Hitler came to power and began passing anti-Jewish laws, and even as his power grew and his violence became more obvious, US Jews weren't sure how to react. Christian pacifists opposed entry into war, as did a fifteen-million-strong group called the America First committee.[100] Of America First, Rabbi Wise said, "We are Americans first, last, and all the time. Nothing else that we are, whether by faith or race or fate, qualifies our Americanism." But in 1938, half of Americans said they had a low opinion of Jews.

In this environment, American Jews were undecided on what to do in response to Hitler. The American Jewish Congress and Jewish War Veterans favored public protest. The American Jewish Committee feared that protest could backfire and get German Jews killed; leaders of the Jewish community in Germany had themselves urged Americans not to protest. The American Jewish Committee and American Jewish Congress, as well as B'nai B'rith, all agreed that a boycott of German goods would only lead to repercussions for German Jews.

"During this election campaign just over," *B'nai B'rith Magazine* read in 1936, "we heard a great deal to this effect: that the Jew efface himself as much as possible from public life lest he appear too prominent and make himself a shining mark for enemies."[101]

It wasn't that American Jews didn't care—many cared very much. "I cannot remember Jewry being so wrought up against anything happening to American Jews as the sudden reversion on the part of a great and cultured and liberty-minded people [the Germans] to practices which may be mildly characterized as medieval," Rabbi Wise said.[102] But some in the Jewish community didn't want to seem too pushy or partisan.[103] And if it were Jews who were seen as advocating for other Jews, it could, they worried, seem like a special Jewish issue that only mattered to their own community; some Jewish leaders were enraged, for example, when in 1938 it was Mannie Celler, the Brooklyn lawmaker who owed his original victory to immigrants in his community and remembered this debt always, who introduced a bill that would have allowed the president to lift immigrant quotas to admit victims of religious, racial, and political persecution. "Celler, to our amazement, is introducing a resolution into Congress . . . which is very bad, so bad that it almost seems the work of an agent provocateur," Wise said, adding, "He should have had a non-Jew introduce the measure."

November of that year was Kristallnacht, a pogrom carried out against Germany's Jews by paramilitary forces. Roosevelt continued to reject proposals to help Europe's Jews.[104] The American public did not want more Jewish immigrants. The American public did not want to go to war for Jews. And Roosevelt abided by their wishes.

Matters did change somewhat after the United States entered the war after the bombing of Pearl Harbor—and after Germany declared war on the United States. We cannot know whether the United

States would have immediately declared war on Hitler's Germany, or if American ambivalence toward war in Europe would have relented, if Hitler had not done so. What we do know is that, once US involvement in World War II started, to stand apart from Nazi Germany, US propaganda sought to make the United States seem like a bastion of equality. In 1943, under the Immigration and Naturalization Services, Jews were no longer categorized as "Hebrews," even though all other new European arrivals were still classified under the "List of Races and Peoples." The government was finally willing, after repeated requests from Jewish organizations, to put space between themselves and the Nazis, who, as Eric Goldstein notes in *The Price of Whiteness*, used this government classification scheme.[105] (It is here worth noting, though, that not everyone benefitted from Uncle Sam's newfound open-mindedness: Japanese people and Japanese Americans were interned for at least some of the same time.)

And then the depiction of Germans in mass media changed. In the documentary *Five Came Back*, William Wyler, an American Jew born in Alsace, recounts that, originally, he had felt pressure from his production company to make the German character in *Mrs. Miniver*, a movie about Britain in the war, more sympathetic. He couldn't, he said. There was one German character, and that German character thus had to represent the Nazis, who were perpetrating evil. It was only after Germany declared war on the United States that the production company suggested that, actually, that characterization wasn't such a problem.[106]

American Jews, then, perceived Germans differently than the rest of the country did leading up to America's own involvement in the war. But, then, Jews had an entirely different conception of why war was being waged than other Americans did. Jews understood the war as being primarily about what was happening in Europe, but many non-Jews considered the war to primarily be about Japan, the country that had attacked *them*.[107] And though Jews did sign up to serve, there remained a stereotype of sickly Jews sending good Christian Americans off to fight for them. A City College professor joked, "The Battle Hymn of the Jews is 'Onward Christian Soldiers, we'll make the uniforms.'" Some claimed, baselessly, that Jews in military service were all stateside, in the "Zone of the Interior."[108]

There were also barriers to Jews joining the military. As Deborah Dash Moore recounted in *GI Jews: How World War II Changed a Generation*, a young man who wanted to sign up for the US Navy in New York was told there was no room for him after saying he had never been in the Boy Scouts, and so signed up in Baltimore, where he appeared to the Naval Reserve office as white, not as another Brooklyn Jew.[109] Once in the military, Jews served with men who had quite literally never met Jews before and withstood antisemitic taunts.[110] Some Jewish men who served in World War II, surrounded by Christians and Christianity, converted.[111] But for others, it was their understanding of Jewishness that changed. To be Jewish and to be American were one, but—particularly for those who served not in Europe but in the Pacific—Jewish identity or understanding of Jewishness became private and internalized. American notions of Jewishness were challenged, too. One young Jewish American, stationed in Calcutta, met Indian Jews and realized how limited American Jews' conception of Jewishness as being tied to whiteness—and even Jewishness as being tied to Yiddishkeit—was. He became a Zionist, convinced that only the Jewish homeland could unite, not divide, the Jewish people.[112]

Antisemitism didn't just play out in Americans' understanding of why they were fighting, and it wasn't limited to the armed services. The United States did not want to accept Jewish refugees. The State Department, inconveniently for the Jews of Europe, was a bastion of antisemitism. Undersecretary of State William Phillips wrote in his diary about a trip to Atlantic City that the place was "infested with Jews. In fact the whole beach scene of Saturday and Sunday afternoon was an extraordinary sight—very little sand to be seen, the whole beach covered by slightly clothed Jews and Jewesses."[113] Some diplomats dismissed the reports Wise got from European Jews on the situation in Germany as "a war rumor inspired by fear." The presentation in the press was muted, too: in 1942, the *New York Times* ran "Himmler Program Kills Polish Jews" on page ten.[114]

In 1944, Treasury Department officials investigated the extent of the State Department's efforts to keep Jewish immigrants from coming to the United States. The subsequent "Report to the Secretary on the Acquiescence of this Government on the Murder of Jews" ended with a quote from Celler: "If men of the temperament and philosophy of [Breckinridge] Long continue in control of immigration administration, we may as well take down that plaque from the Statue of Liberty and black out the 'lamp beside the golden door'"[115] (a reference to a poem below the statue, written for a contest by Emma Lazarus, who descended from Ashkenazic and Sephardic immigrants).[116]

In fairness to the antisemites in charge of the country's foreign affairs, the State Department was not the only government arm that would not budge. Congress, too, refused to help the Jews of Europe. In 1939, the *St. Louis*, a ship with nine hundred passengers, the majority of whom were Jewish refugees, tried to let its passengers disembark in Cuba; they were turned back. Some were taken by other countries in Europe; an estimated 254 died, many in concentration camps. That same year, Congress fought a law that would have permitted the entry of twenty thousand refugee children more than the immigration quotas allowed. Opponents didn't like the idea of young Jews coming into the United States when, per North Carolina Democrat Robert Reynolds, refugees were already "systematically building a Jewish empire in this country." The bill never even reached the Senate floor.[117]

Half of Americans in 1943 refused to believe reports that the Nazis were systematically murdering Europe's Jews. This led Jewish

Hollywood screenwriter Ben Hecht (who, in another illustration of the complication of early-twentieth-century Black-Jewish relations, was an author of the screenplay for *Gone with the Wind*) to write *We Will Never Die*. Hecht brought Jewish celebrities (and Frank Sinatra) together to act out the plight of Europe's Jews in a stage show that opened in Madison Square Garden and then toured the country. At the end of it all, Hecht concluded that he had "accomplished nothing."[118]

And then, of course, there was Roosevelt himself, who, even after Kristallnacht, said that relaxation of immigration restriction was not under consideration. In 1943, five hundred Orthodox rabbis marched on Washington pleading for more support for those European Jews who were still alive. But the president didn't meet them.[119]

Discretion didn't work; pleading didn't work; marching didn't work. The United States did not want to save Europe's Jews. It didn't want to go to war for them, and it didn't want to let them in.

According to opinion polls, in 1944, antisemitism in the United States was at the highest point it had ever been.[120]

Wyler—the director who was allowed to portray Nazis negatively in film after Germany declared war on the United States—was sent to the western front to make propaganda films. When he returned, he heard a man outside a hotel make an antisemitic comment. He assaulted the man (and so was arrested). He was confused, and hurt, because he had thought he had gone to war and served for the United States to fight antisemitism, but here, at home, he found himself confronted by it.[121]

But many Americans did not feel conflicted about hating Jews and fighting Nazis. "I don't think most Americans saw fighting Nazi Germany as fighting for Jews," Kenneth Jacobson, deputy national director of the Anti-Defamation League, told me over the phone. "It was about fighting a group and a country that was committed to destroying other people, including us.

"I'm not sure that there was a connection in most people's minds to protecting Jews," Jacobson said.[122]

Antisemitism did indeed decrease after World War II. But by then, millions of European Jews had been murdered, and American Jews had been changed forever.

"The burden is solely ours to carry. Jewish culture and civilization and leadership are shifting rapidly to these shores," said Hebrew Union College historian Jacob Rader Marcus.[123]

Before the Holocaust, American Jews had grappled with how to be Jewish, and what it meant to be Jewish, and indeed even with what a Jew was. But now all those questions were, in a way, more urgent. There were several million fewer Jews in the world. Over the course of just a few years, the proportion of the world Jewish population in the United States suddenly and dramatically increased. American Jews hadn't asked for that, or sought that out, but it had come to them.

"American Jews are realizing that they have been spared for a sacred task—to preserve Judaism and its cultural, social, and moral values," read the *American Jewish Yearbook* of 1941.[124]

This burden was felt in different ways. Intermarriage, of course, came under attack again. The wife of a leading Conservative rabbi by the name of Rose B. Goldstein directly connected intermarriage to the continuity of American Jews and, by extension, Jewishness. "The future of American Jewry," she said, "is directly conditioned on the education of its womanhood."[125]

Many American Jews also became reinvigorated in their own faith during and after World War II. "Jews . . . who had abandoned their people" were "returning like prodigal sons," Mordecai Kaplan, the father of Reconstructionist Judaism, said.[126]

This idea—that American Jews now understood themselves to be tasked with preserving Jewishness—was true even of secular Jews.

The Yiddish poet Jacob Glatstein wrote, "I'm going back to the ghetto . . . I cry with the joy of coming back." And in 1938, Sholem Aleichem Folk Institute leaders, which were previously fiercely secular, decided "to introduce the study of Pentateuch into the elementary schools, to emphasize the celebration of Jewish holidays, and, in general, to establish a positive attitude towards all manner of Jewish ways of life."[127]

Jews in America had, in a way, outrun history. They had the chance to do what millions of European Jews had not: to continue to live as Jews. But what did it mean to live both a fully Jewish and fully American life?

For my nana, it meant an insistence on supplemental Jewish education and strong resistance to intermarriage. For my grandfather, it meant becoming an intense supporter of Israel.[128] These were, as it turns out, two very common—and fraught—responses.

Chapter 2

White and Red Jews

I N 1959, AMERICAN BUSINESSWOMAN Ruth Handler, along with her husband, Elliot, the cofounder of Mattel, invented the Barbie Doll. Barbie—blond and white and with proportions that made her more all-American than any living American woman could naturally be—was named for Ruth's daughter, Barbara Handler.[1] The story goes that Ruth noticed Barbara was playing with paper dolls as though the dolls were not babies but adult women in some future time and place. Their paper clothes and paper form were not up to the task.

There is a certain irony in a Jewish woman inventing this icon of white America. Legally, Jews of European descent have always been considered white in the United States. But culturally, this status has been more in flux. Before World War II, Americans, Jewish and non-Jewish, struggled with where to pin Jews in America's racial hierarchy. In the years from 1945 through the 1960s, many American Jews moved more comfortably into and up in the world of white America. Some Jews were ambivalent about whiteness; others actively tried to position themselves as white, and specifically as white in the context of the Cold War—and in opposition to Black Americans.

Barbie's Dream House first came out in 1962.[2] In real life, the American Jewish dream house was defined not only by those whom it ostensibly sheltered but also by those who were kept from living under its fantasy roof.

The whiteness of Jews in the United States has been taken as obvious, then questioned, then reasserted over the decades.

Before the mid-nineteenth century, European immigrants who came over generally assimilated into white America. This was no small thing. In the American context, to be white was to be in the part of society that traded and owned and profited from enslaved people, and not the part that was enslaved. That included European Jews, who were broadly thought of as white.

Elsewhere in the Atlantic world this was untrue. There were other parts of the western hemisphere where Jews were not predominantly white. There was, for example, Suriname, where Portuguese Jews and their descendants had autonomy within a slave society. Those Jewish men who came over from Europe to live and work in Suriname raped the women they enslaved and then had their children converted to Judaism. In this way, the "Eurafrican" Jewish population likely came to surpass the "white" Jewish population there by the early 1800s.[3]

But the United States was not Suriname. And in the United States, Jewish whiteness allowed some Jews to play an active role in upholding America's racist, slave-based society. In chapter one, we learned of Judah P. Benjamin, but he was hardly alone. The first Jewish member of Congress, Lewis Charles Levin, elected in 1845, founded the American Republic Party, later known as the Native American Party—and informally known as the nativist Know Nothing Party.[4]

Other American Jews also fought on the side of the Confederacy during the Civil War, making arguments that Judaism and slave ownership were coherent.[5] That many Jews fought on the side of the Union and made the opposite case arguably had less to do with some inherent Jewish value and more to do with the fact that more Jews lived in the North.

Jews in the United States before the mass Jewish migration also spoke about themselves as members of a religion, not as a distinct people or a race: consider, again, the Pittsburgh Platform of 1885, which said explicitly that to be Jewish was to belong to a faith, not a nation.

But as more Jews came to America, Jews of European descent were to be seen as something other than white. After 1880, immigration shifted to become primarily from southern and eastern Europe and from less wealthy communities, changing the makeup of cities. Immigrants and their children made up 70 percent of America's largest cities, which in turn assumed a southern and eastern European character. Immigrants were not just disappearing into white American culture; they were changing what it looked like. This caused consternation for those white Americans who were already here who did not want the portrait of white America to change.

There are several theories as to why Jewish racial status was thrown into question in the late 1800s. Karen Brodkin, in her *How Jews Became White Folks and What That Says About Race in America*, writes of how Jews were concentrated in the garment industry in part because they were frozen out of printing, carpentry, painting, building, and highly unionized fields like transportation and communication. The American Federation of Labor was, at the time, adamant that the "privileged labor class" that needed to be protected was white America of German, British, and Irish descent. "In sum, the temporary darkening of Jews and other European immigrants during the period when they formed the core of the industrial working class clearly illustrates the linkages between degraded and driven jobs and nonwhite racial status," Brodkin writes.[6] Brodkin was referring to the late nineteenth and early twentieth centuries, linking American Jews' working-class status to their status as "lesser than" in the eyes of some of their countrymen—though the fact that Jews were considered to be less desirable Europeans did not actually change their racial legal status in the United States.

Further, while many Jews worked in what were viewed as lower laboring classes, which Brodkin argues changed the perception of Jews for the worse, white America was *also* uncomfortable with how many Jews were coming into positions of status. Jews, who treated universities not as clubs for privileged play but opportunities for self-advancement, did so well in the highest echelons of American education that quotas were imposed at many of the most prestigious universities because these institutions were first and foremost places where white Americans could secure positions of prestige and power in white America.[7]

This is what works like *The Melting Pot* missed. It was not only that newer Jewish immigrants were resistant to let go of their culture or to see themselves as just another set of white Americans, and it wasn't just that Jews insisted that intermarriage was a problem for the continuation of Jews and Judaism. America demanded assimilation, but, as more southern and eastern European immigrants arrived, Americans decided that they also had strong beliefs about what sorts of people were capable of assimilating.

In fairness to mainstream white America, newer, less acculturated immigrants were also unsure that they *wanted* to be a part of mainstream white America, at least upon arrival.

Jews who had been in the United States for longer, or whose families had long been in the United States, understood very clearly that whiteness was power and safety.

More recent eastern European Jewish immigrants felt differently about all of this than their more established peers. Freshly arrived and coming from a place with its own violent set of prejudices, they felt less of a need to self-define in a way that met white America's expectations. They had their own sense of what it meant to be Jewish, and it wasn't just "people who prayed differently." In fact, even the concept of race-based divisions didn't quite fit in the eastern European Jewish mindset: a survey of Yiddish press and popular culture suggests that they spoke of *dos yidishe folk*, or Jewish people, and not *di yidishe rase*, or Jewish race. In the United States, race was the organizing principle, but that hadn't been the case in eastern Europe, where Jews thought of themselves as a people or ethnicity, but not a race, apart.[8] Was the question of racial sameness or distinctiveness, some wondered, even relevant to them?

There were whole communities that revolved around the Yiddish language and Yiddish culture, some of which, like the Workmen's Circle, were also Socialist. These pockets were insular but not isolated. For one thing, over time the prevalence and power of Socialist Jews—or of Socialism in Yiddishkeit—decreased. According to Arthur Liebman, in his *Jews and the Left*, it was because these groups tried to work with other (non-leftist) Jewish organizations and communal structures. They could not afford not to do so. But in doing so, he argues, they became more Jewish—or, rather, more of a different kind of Jewish—and less leftist.[9] But there was another dynamic at play, too: even Yiddish-language newspapers could not keep Jewish communities away forever from the predominant way of thinking of hierarchy in the United States, which was, from its founding, along lines of Black and white.

Some Jews sought to prove their whiteness by denigrating non-white—and specifically Black—Americans. As Eric Goldstein examines in *The Price of Whiteness*, in some urban Jewish neighborhoods, lower-middle-class Jewish women would bargain with Black women for their services, offering somewhere between 15 and 35 cents an hour. They couldn't afford luxury servants, but they could denigrate Black women's labor at what came to be known as "slave markets."[10] Similarly, some Jews tried to push back against the popular race science of the day that said Jews weren't white with even more race science. Maurice Fishberg, a leading scholar of Jewish physical anthropology, took skull measurements. He argued at first that being "Semitic" couldn't possibly mean being "African" and that the Jewish skull type meant that "Semites" could not have originated in Africa. This was broadly unsatisfying, since the link between the Jewish people and Africa was well accepted by this point. Fishberg moved on to study Jewish pigmentation and argued that Jews' light skin as well as their hair and eye color meant that Jews couldn't be of purely Semitic or African descent. He then compared Middle Eastern Jews to Jews of European descent—that is, the majority of Jews in America at the time—and concluded that European Jews had been so dramatically transformed through intermarriage that they had "no relation at all" to those who bore a closer resemblance to Africans. In 1911, he published *The Jews: A Study of Race and Environment*. "It is clear that certain strata of the population cannot assimilate merely by adopting the language, religion, customs, and habits of the dominant race," wrote Fishberg in a later article. But "the Jews, as whites, are by no means debarred from assimilating with their fellow men of other faiths."

Fishberg accepted the racist language of eugenics and left no room for Jewish distinctiveness, and so did not satisfy most American Jews, who wanted both white American acceptance and a distinctive identity.[11]

Opposition to this line of thinking did arise from other Jews, though many who pushed back were unaffiliated with organized religion. Jewish Communists, for example, had in theory broken with the Jewish community, but their critique of capitalism was coming from an unmistakably Jewish tradition: the Bund, after all, was explicitly Socialist. Writing in *The Liberator*, Michael Gold penned, "I have no sympathy for the upper-class Jew who wails against discrimination at Harvard or at high class hotels, because all of them unite in discrimination against the Negro." In Harlem, Jewish Communists led local Black Americans in rent strikes against their Jewish landlords and joined in boycotts against Jewish merchants. They also built one of the city's first racially integrated cooperative housing projects, right by one of the aforementioned "slave markets" where Jewish women bid for Black labor at low prices. According to one account, the Black children in the housing project spoke Yiddish better than their Jewish playmates did.[12]

Franz Boas, an anthropologist of Jewish origin who worked to discredit race in evaluating intellectual capabilities, identified usually as a German American, and not as a Jew.[13] In the early 1900s, the National Association for the Advancement of Colored People was organized and established; social worker Henry Moskowitz, who was Jewish, was heavily involved early on.[14] Later NAACP presidents, like Joel and Arthur Spingarn, were also Jewish but had no formal connections to the Jewish institutional world.[15] Many Jews who were involved in or supportive of Black-Jewish relations or civil rights were primarily connected to or came from the Yiddish world. The Yiddish press was particularly enthusiastic in its support for the civil rights movement. As Yiddish writer Isaac Rontch put it, the Yiddish writer, "himself a child of an oppressed people, is forever on the alert to hear and sense the ever-present protest of the Negro against his white discriminator."[16]

But for every Rontch, there was a Jewish leader arguing that Jews should be counted as white for immigration purposes, or a Jewish woman seeing how little money she could offer to a Black woman to clean, or a Jewish shop owner discriminating against Black customers. Whiteness, malleable though it is, has a constant, which is that it must always be defined in opposition to something. Precisely because race is a construct, the boundary between "white" and "not white" is slippery and inconsistent. And so there was a perceived benefit, if not tangible then emotional, to these American Jews discriminating against other immigrants and against Black laborers and customers. So long as they could discriminate, they could prove, if only to themselves, which side of the boundary they were on. They could prove that they were not *not* white.

American Jews were aware of how precarious whiteness made status and how easily those who tried to cling to it could be cast out. Consider that in 1910, there were plans for the census to include immigrant "races." This was part of a broader trend; public schools in Philadelphia had started asking for their students' racial backgrounds. Those Jewish students who simply said they were "American" were asked to correct their entries.[17]

American Jews, or at least many of their families, knew persecution. They knew that American whiteness meant shedding distinctiveness: your language, your traditions, and in some cases your political values. American Jews knew all of this. But they, or at least many of them, were also choosing to move closer to the white American mainstream. It was the white mainstream that promised stability and safety. This was, of course, a promise that could easily be broken, and that, in any event, was built on taking stability and safety from others. But it was a tempting promise, all the same.

"Despite their desire to find social acceptance while also preserving a distinctive minority sensibility," Goldstein writes, "most Jews of this period found that these two goals could not be easily harmonized."[18]

This tension was about to be exacerbated further.

As American Jews gained a little more financial and status stability, many moved out of ethnic enclaves. But the great move into mainstream white America—into the dream house—came with World

War II. Jewish Americans, in benefitting from the GI Bill, enshrined their status as white Americans.

Recipients of the GI Bill were the most rewarded military veterans in US history.[19] The GI Bill offered money for training and education; loan guarantees for homes and businesses; and unemployment pay. Almost half of World War II veterans had received education or vocational training by the time the original bill expired. In 1955, the Veterans Administration backed 4.3 million home loans. The 1950s—the white picket fences, the moves to suburbia, the leisure time, the move of millions into the middle and even upper middle class—were made possible by the GI Bill. Children of Jewish immigrants who served in the war who may not otherwise have gotten to go to college or buy homes were financially empowered to do exactly that.

But while the GI Bill was race neutral in theory, its House sponsor, well-known racist representative John Rankin, made sure that it would not be race neutral in practice. Rankin included the words, "No department, agency, or officer of the United States, in carrying out the provisions of this part, shall exercise any supervision or control, whatsoever, over any State educational agency, or State apprenticeship agency, or any educational or training institution," which effectively meant that, particularly in the South, where segregation was both the law and violently enforced social practice, Black Americans were not able to receive the benefits that they were due under the GI Bill. A Black veteran couldn't receive an education if the college did not accept Black students; a Black veteran couldn't get a home loan because the Federal Housing Administration, charged with financing postwar housing developments, prohibited homes from being sold to Black Americans.[20]

This is just one example of the ways in which Jewish whiteness, relative to Blackness, repositioned Jews in the postwar period.

Two trends accompanied American Jews' postwar middle-class ascent. The first was a move into suburbia. The second was a renewed understanding of Jewishness as, first and foremost, a religious, and not ethnic, identity.

American Jews may have flocked to the suburbs faster than other Americans at the time. In New York, American Jews made their way to the suburbs of Long Island. Between 1940 and 1960, the population of Long Island's Nassau County grew 220 percent, but its Jewish population grew 1,770 percent.[21] According to the US Census, more Americans did not live in the suburbs than the cities until the 1970s. But in the Cleveland area in 1958, 85 percent of Jewish Americans lived beyond the city limits. In becoming suburban, Jews also moved beyond the places in which they'd previously been concentrated and to other parts of the country, notably Florida and California. And it took their institutions out of cities, too. In 1969, the Jewish Community Center of Washington, DC, moved from the nation's capital, where it had stood close to the White House since the 1920s, to Rockville, Maryland.[22]

The very act of moving reinforced the whiteness of American Jews. In 1952, Brooklyn-based associate rabbi Benjamin Kreitman argued that the demographics of his neighborhood changed as dramatically as they did because the Jews who had lived there were not serious in practice about the principles of equality and integration.

Rabbi Israel Leventhal of the Brooklyn Jewish Center similarly said he felt "indignation" when Jews moved out to avoid having Black neighbors. He allowed that there were some who had "valid" reasons for moving to suburbs—that it was fine if a move ended up with segregation so long as it was because nice Jewish parents wanted their children to have a lawn and good public schools, just as long as they weren't moving to those neighborhoods because Black people couldn't. But as Rachel Kranson writes in *Ambivalent Embrace*, "liberal Jews could not always tell where upward mobility ended and racism began." To move up in the United States was to participate in its racist systems.[23]

• • •

My father, born in 1959, a child of this Golden Age of American Jewishness, told me that, growing up, Jewishness "was always something all around us. It was very important in my family. My mother was all about it." But my nana was all about it, to use his parlance, in a very particular way.

"Temple was all important," my father told me. "The temple community was all important." If temple was 1a, then 1b, he said, was teaching her children Jewish history and making sure that they learned Hebrew and the religion (though, a public-school teacher herself, she considered Jewish education a supplement, not a replacement). But 1a was the temple. "That was everything to her. And going to, making sure that we went to synagogue every Saturday, come hell or high water, *blizzard* or not, it was huge to her."

"Most of our family social stuff surrounded that temple," he said. "It was a huge element in my early life. Huge." He added, "The Sabbath was fuck-all boring." They walked to their Conservative synagogue (no riding in cars on Shabbat), ate knishes and similar kosher fare, and his parents rested (no work allowed on the Sabbath). "That's how shocking it was that, when I cut my face open (on a Saturday), my mother actually got in a car to go with me to the hospital."[24]

My nana was not alone in her singular devotion to synagogue, or in revolving her life and her family's understanding of Jewishness around it.

Decades earlier, when American Jews primarily lived in a handful of cities, Jewishness was everywhere. One could abstain from going to temple, but one would still be hit with the book stalls and the language and the food. The Jewishness on those city streets was, of course, one particular, Ashkenazic-centered version of Jewishness, but these communities had *a* version of Jewishness that overwhelmed the character of the streets where the country's Jews lived.

This was not necessarily true of the roads and lanes out in the suburbs. The streets there had something, though: they had synagogues.
Synagogues became the site of postwar Jewishness because American Jews were taking part in a national "religious revival." Church
and synagogue membership grew by over 30 percent in the 1950s.
In 1954, President Dwight Eisenhower made the case for adding
the words "under God" to the pledge of allegiance, citing "spiritual
weapons" that could be used against America's atheist Cold War enemy, the Soviet Union; the change was signed into law.[25]

Jews did not actually actively go to services at the same rate that
Christians did: in 1958, only 18 percent of Jews attended weekly
services, compared to 76 percent of Catholics and 40 percent of
Protestants.[26]

But Jews did *join* synagogues. In the late 1950s, 60 percent of American Jews—over three million—belonged to a temple. It was the first
time since colonial days that over half of America's Jews elected to
affiliate.[27] To have membership in a synagogue was a way for suburban Jews to signify that they were a part of the Jewish community; it
was a way, out away from Yiddish newspapers and crowded streets,
to show to others, and perhaps even to themselves, that they were
still Jewish.

And American Jews weren't only joining synagogues: they were
also building new ones. In 1956 and 1957 alone, the United States
saw the establishment of twenty new modern Orthodox, twenty new
Reform, and forty-one new Conservative synagogues.[28] Conservative
Judaism did particularly well in this period of American Jewish history, a happy midway for the suburbs between Orthodoxy, seen by
many as incompatible with American suburban life, and Reform Judaism, which was seen as not Jewish enough, and which started to
reintroduce religious rituals in an attempt to draw people back in.[29]

One of those rituals—arguably *the* ritual of the American Jews'
suburban experience—was the bar and bat mitzvah, the Jewish

coming-of-age ritual. The bat mitzvah, which is for girls, was a newer tradition: the first in the United States is thought to have taken place in 1922, when Mordecai Kaplan performed one for his daughter, Judith.[30]

But it was in the postwar period and in suburbia that the bar and bat mitzvah became more than a religious ritual. It became, very often, a party and a show of the family's wealth and connectedness. Some in the Reform movement, in fact, pointed to the bar and bat mitzvah as proof that they were right to have done away with the rituals that were now coming back. In 1959, Rabbi Joshua Trachtenberg of New Jersey explicitly told *Time* that they typified problems with the Reform movement's return to ritualism and dismissed them as "empty ceremonialism," characterized by "conspicuous waste."[31]

In fairness, it is not as though money was only being spent on preteens' parties. America's Jews spent somewhere between $500 and $600 million on new synagogue construction between 1945 and 1950.[32]

The 1950s and '60s were proof that, if you build it, they will, if not come weekly, then pay membership fees and come on the High Holy Days (and some synagogues were built with this in mind, complete with sliding walls that could make a room small or vast depending on whether it was a regular Shabbat or a High Holiday).[33]

Synagogues became the site of Jewishness, the sun around which my nana's Jewish world orbited, in part because religious affiliation, American Jews had learned, was comprehensible to the rest of America in a way that a secular Jewish identity was not.

In 1955, Will Herberg, a Jewish philosopher and former Marxist, published *Protestant, Catholic, Jew*, a book that presented being Jewish as being about religious, not ethnic, difference.[34] The United States, from the beginning, understood Black and white and religious freedom: it was less comfortable with concepts like ethnicity and peoplehood and groups that transcend race. With synagogues,

American Jews could be publicly, outwardly all-American and privately, religiously all-Jewish. They could move out to the suburbs.
They could move up the socioeconomic ladder like other white
American families—they would just be white American families who
happened to go to shul. It was, in a way, a return to the way Jewish
identity was understood before there was an increase in immigration: Jews could be white Americans who prayed differently. This was
reflected in how Jews were depicted on television. When Gertrude
Berg's radio broadcast about a Jewish suburban family became a television show, the only signifier that the family was Jewish was a menorah on the mantelpiece.[35] This wasn't threatening to white America.
This was acceptable and comprehensible.

It helped, too, that this was during the Cold War, with the United
States angling to prove it was both full of faith and tolerant of all
religions. In a way, this was a sort of mirror image of what the Soviet
Union was doing: in the Soviet Union, religion was not tolerated,
but propaganda films were made to show how racially accepting the
Soviet Union was in comparison to the United States. The United
States, meanwhile, wanted to show itself as a land of religious tolerance while continuing de facto and de jure racial segregation. This
was part of the reason that Judaism was not only understood by
Americans but was increasingly understood as *being* American, considered a major religion in this country despite the fact that fewer
than 5 percent of Americans were Jews.[36]

The idea that American Jews were keepers of this faith, which developed after the trauma of the Holocaust, was also part of the new
conception of Judaism in America.

As historian Hasia Diner has shown, and counter to some popular historical narratives, the Holocaust was a trauma that was understood and processed in the immediate postwar period. Though the
word "Holocaust" was not universally used until the 1960s, as early as
1953, historian Rufus Learsi created a reading of remembrance that

was distributed nationally and locally at Jewish community councils, Jewish press, and synagogues. It was meant for use during Passover seders and linked the story of Passover to the uprising in the Warsaw Ghetto of 1943. In 1948, the Reconstructionist movement's mahzor included prayers for murdered Jews of Europe and specifically cited the evils perpetrated by the Nazis. Jewish summer camp seized on Tisha B'Av, an annual fast day to commemorate the disasters that have befallen the Jewish people and specifically the destruction of the Temple in Jerusalem, to remember the devastation of the Holocaust. American politicians were attending Holocaust remembrance events by the early 1960s, acknowledging by doing so that there were people in their midst who were impacted by it.[37]

But if the postwar, Cold War period made the United States embrace Jewish religious institutions, it also made the United States less tolerant of some other Jewish cultural institutions. This was the other reason that synagogues were what Jews brought with them to the suburbs. Synagogues weren't suspicious. Between disappointment in the ostensibly leftist empire—the Soviet Union (and Communist China)—and America's increasingly hostile climate to leftist politics, which disproportionately affected Jews, who were already suspected of being Communist sympathizers, support for leftist institutions dwindled. Jews, as we saw in chapter one, did make up a disproportionate number of Socialists and Communists, leading some to believe that Jews were all radical revolutionaries lying in wait to disrupt wholesome society.

American authorities sprang into action to stamp out this radicalism. Senator Joseph McCarthy was singularly dedicated to ferreting out Soviet sympathizers from American life. Fifty-six percent of Catholics and 45 percent of Protestants approved of McCarthy's mission; only 2 percent of America's Jews did.[38] Efforts by McCarthy himself tended to target Protestants, not Jews, but evidence suggests that anti-Communist investigations at the state and federal level disproportionately focused on Jews.[39] As early as 1941, Senators Burton Wheeler and Gerald Nye investigated Hollywood's role in promoting Soviet propaganda. Wendell Willkie, an early advocate of rescuing Jews from the Nazis and creating a Jewish state, was an attorney working on behalf of Hollywood's studios. He argued that, by "Communism," the senators meant "Judaism," and in so doing tried to paint the investigators not as patriots but as antisemites. But the accusation of antisemitism didn't prevent the House Un-American Activities Committee from investigating Soviet influence in Hollywood. Ronald Reagan, then president of the Screen Actors Guild, made an early name for himself testifying before HUAC. The committee did not favor everyone as it did Reagan, who would be elected US president a few decades later: the ten writers (six of whom were Jews) who refused to cooperate with Congress, known as the Hollywood Ten, were, for their principles, blacklisted.[40]

This was obviously significant: for standing by their principles, ten writers were cut off from their livelihood. Even so, the single most high-profile case did not involve the silver screen. It was that of Julius and Ethel Rosenberg.

The Rosenbergs were a husband-and-wife pair, members of the Communist Party, who were tried in 1951 for spying for the Soviet Union. They were accused of providing information about radar and sonar, but also on the designs of nuclear weapons. David Greenglass, Ethel's brother, worked at Los Alamos, where America's atomic bombs were being developed. He and his wife, Ruth, were also charged, but these charges were dropped after Ethel's brother said that she, Ethel, typed up the notes.

Both the American Jewish Committee and the Anti-Defamation League did their own purge of Communists and Communist sympathizers. The American Jewish Committee turned files over to the House Un-American Activities Committee, which was charged with finding Communist threats to the nation. In the Rosenberg case,

HUAC went further. An American Jewish Committee staffer wrote a pamphlet, *The Rosenberg Case*, that supported the death sentence, and historian Lucy Dawidowicz wrote in *Commentary* that the couple should be executed; clemency, she wrote, would be bad for the Jews.[41]

Robert Meeropol, one of the Rosenbergs' sons, now grown, told me in a 2020 interview, "Over the years, I've spent a lot of time . . . looking into the reaction of the Jewish community and trying to understand it."

The National Committee to Secure Justice in the Rosenberg Case, he said, was mostly made up of Jews. And Orthodox Jews "were appalled by the death sentence and had a better recognition than the mainstream Jewish community that antisemitism was involved." But mainstream Jewish organizations and Judge Irving Kaufman "felt that if they wrapped themselves in the American flag, they could prove their loyalty to America, by not coddling their co-religionists."[42]

There are some who believe Jewish organizations took this position because they were afraid. But historian Hasia Diner told me, "The context as I understand it of that behavior was [a] long ongoing internal Jewish war against the Communists."

That isn't to say that fear wasn't involved at all. But mainstream Jewish organizations "didn't like Communists, didn't like behavior within communal institutions, disagreed with the philosophy and so on. To me that long struggle is as much the context as fear of antisemitism."

The whole thrust of American political life at the time, Diner said, was about how Communists were evil, which "made it very dicey for organizations like the American Jewish Committee, which saw itself as *the* voice, *the* defender of the Jews, to constantly live in this environment where Jews and Communism elided in public discourse." Still, they could have said that they didn't like Communism but they didn't believe that execution was appropriate. And they didn't. Jewish organizations and institutions, consciously or not, took an opportunity to distance themselves from Communism. And while there were some individual Jews who stood up for and by the Rosenbergs, the vast majority kept their heads down.

"I don't think it was American Jewry's finest hour," Diner said.[43]

The Rosenbergs were found guilty, and in 1953 they were executed.

In 1995, the Venona files, decrypted messages sent by Soviet intelligence agencies, were made public. The files showed that Julius committed espionage but suggested that Ethel was only an accessory.[44]

Later, David Greenglass would recant his testimony and say that, in fact, he couldn't remember who typed up the notes and had only pointed to Ethel to spare his wife. It was probably his wife, Ruth Greenglass, who had typed up the notes. In journalist Sam Roberts's 2001 book, *The Brother: The Untold Story of Atomic Spy David Greenglass and How He Sent His Sister, Ethel Rosenberg, to the Electric Chair*, Greenglass was quoted as saying, "My wife is more important to me than my sister. Or my mother or my father, O.K.? And she was the mother of my children."[45]

In 2008, codefendant Morton Sobell spoke out to say that the military secrets he and Julius Rosenberg passed along were confirming what the Soviets already knew, and that Ethel Rosenberg had no active role.[46]

But by that time Ethel Rosenberg had been dead for over fifty years. Her sons tried to get the Obama administration to exonerate her; Obama did not. They did not bother in the Trump era.[47] Trump, after all, was connected to Roy Cohn,[48] one of McCarthy's two partners in ferreting out Communist influence both real and imagined.

The story of Ethel Rosenberg is, in many ways, a Jewish story. Or, rather, it is a series of Jewish stories, linking together and challenging one another. And looking at them, one could ask: Whose is the most authentically Jewish story? Who, in this, is the Good Jew or the Bad Jew? Is it the Communist Jewish woman who was executed? The Jews who stood by her? Or the Jews who called for her death? The sons, years later trying to push for the exoneration of their mother? Or Roy Cohn, the Jewish man who helped create the environment that killed her? Which one of them gets to claim the largest place in Jewish history? Or do they all sit there together, sharing the mantle of "Jewish" together now as they did not then?

Leftist institutions also declined because their raison d'être no longer matched American Jewish reality. These were institutions that were formed to serve the working class, and many American Jews were rising through the socioeconomic ranks. Arbeter Ring, or Workmen's Circle, struggled, stuck between whether to rebrand as a group dedicated to universal rights or to continue to organize around labor rights, despite the fact that most American Jews were now middle, not working, class. Sholem Aleichem schools and camps, dedicated to Yiddish language and culture, dwindled in popularity, too.[49] They didn't lose their mystique or appeal to some: Jim Hoberman, *Village Voice* film critic, recounted thinking of Hebrew school as full of privileged middle-class kids, whereas his friend's Sholem Aleichem education seemed like a more "authentic" experience.[50] But they were, to many Jews, of and for a different time and place in Jewish American history.

Jim Hoberman was not alone in his conviction that the way in which he and his colleagues were learning to be Jewish was somehow less authentic, or even wholly inauthentic. Angst over whether Jews, in earning more money and living more comfortably, had somehow ceased to really be Jews was one of the main themes of the day.

"The nearer to temple, the farther from God," the Yiddish saying goes, and this was the sort of thinking over which American suburban Jews agonized at this time. They had built themselves these religious buildings, they were paying the fees, but they did not feel that they had constructed for themselves authentically Jewish lives. John, who described himself as a midwestern Jew born in 1959, said that he always romanticized the childhood of his father, who grew up in Poland. I asked him if this was because his father's life was different from what seemed to him his ordinary American childhood, or if he romanticized his father's youth because it seemed more authentic than his own. Both, he told me.[51]

In addition to castigating themselves and their fellow American Jews over suburban synagogues, some American Jews romanticized the past and the far away. There was a way and a place to be Jewish. It just wasn't here and now.

As Kranson explains in *Ambivalent Embrace*, the shtetl—which literally means "town" or "village" but is sometimes used as shorthand to describe the Ashkenazic Jewish old country, and specifically the parts of eastern Europe in which Jews lived—had been romanticized since the nineteenth century. But the romanticization kicked into overdrive in the postwar period.[52] Sholem Aleichem, the most famous Yiddish-language author, was, in real life, Solomon Rabinovich, who speculated on stocks and married the daughter of a wealthy landowner.[53] Yet his village characters captured Americans' imaginations of what it was like in the shtetl; *Fiddler on the Roof*, a Broadway musical loosely based on Aleichem's stories of Tevye the milkman, premiered in 1964. It depicted Jews in the village of Anatevka, and Tevye belts out songs like, "Tradition" and "If I Were a Rich Man." The show was a smash hit.

In Roman Vishniac's 1947 book, *Polish Jews: A Pictorial Record*, which was sponsored by a Jewish relief organization, the Joint Distribution Committee, the photographer documented life before the Holocaust, showcasing the most pious and impoverished the shtetl had to offer. In fact, Vishniac had also taken photos of cosmopolitan, well-off Jews, but those that did not fit the vision of Jewish life advanced by the book did not make it onto the page.[54]

Similarly, American Jews' earlier years on the Lower East Side were held up as an exemplar of authentic Jewish life. Nathan Glazer and Daniel Patrick Moynihan, authors of *Beyond the Melting Pot*, described the period during which American Jews were packed into the Lower East Side as "When the Jews were thus most Jewish."[55]

But what did "true Jews" or "most Jewish" actually mean? What made one Jewish life more or less authentic than another? Was the Jew who eschewed the synagogue but read a Yiddish newspaper more Jewish than the Jew in the suburb who paid for his daughter's bat mitzvah? What about Jews with money in the shtetl? Were they somehow less Jewish than their poorer neighbors? What's more, not all Jews came from the shtetl or lived on the Lower East Side. The Judeo-Arabic community in Brooklyn was roughly 15,000 strong in 1950. There were approximately 75,000 Sephardic Jews in America in 1934 (though by the third generation, most Sephardim had intermarried).[56] Were Vishniac and Nathan Glazer really arguing that a person who didn't live in the shtetl or who hadn't passed through a certain New York neighborhood wasn't Jewish?

And why, if mainstream white America was not Jewish, was Nathan Glazer actively arguing that it was where Jews belonged?

The majority of American Jews, and certainly the majority of American Jews in the 1950s and '60s, considered themselves liberals. This reflected, for some, a natural evolution of American Jewish politics, from Socialist leaning in eastern Europe and the Lower East Side to liberalism and the Democratic Party in the suburbs; for others, it reflected the belief in a historical connection to the civil rights movement; and for others still, it reflected an appreciation that pluralism was what kept any minority group safe. That does not mean that these American Jews weren't hypocritical or that they always lived their stated values. But it does mean that most Jews, even as they acculturated and amassed greater wealth and security, did not start voting for Republicans.

Still, in the postwar period, certain Jewish intellectuals compared themselves to Black men, using the not-so-subtle argument that they could assimilate into white society and Black men could not. In his 1963 work with Moynihan, *Beyond the Melting Pot*, Glazer compares Jews, a good model minority, with bad "Negros" and Puerto Ricans and less good Italians. In the essay "My Negro Problem—And Ours," which was published in *Commentary*, the American Jewish Committee's publication wing, that same year, Norman Podhoretz wrote, "For me as a child the life lived [by Black boys] seemed the very embodiment of the values of the street—free, independent, reckless, brave, masculine, erotic." Jewish boys, in Podhoretz's depiction, lived in good homes. They had strong mothers who made them hot lunches and sent them off to school in uncomfortable woolen hats. By comparison, Black boys "roamed around during lunch hour, munching on candy bars."

In the prewar period, Karen Brodkin writes in *How Jews Became White Folks and What That Says About Race in America*, leading Jewish intellectuals had tried to keep distance between themselves and bourgeois, white, predominantly Protestant America. Now, though, Glazer and Podhoretz and their set of intellectuals were making a different case: that they belonged to white America, and one only needed to look at Black America to see that that was true.[57]

Glazer and Moynihan predicted a "wane in liberalism" as Jews ascended the socioeconomic ladder. But as Kranson explains, that did not actually come to pass. There was no great conservative shift among American Jews. The towns to which Jews moved in the suburbs of Long Island, for example, became towns with Democratic leadership and towns in which Democrats could be expected to win in presidential elections.[58]

In 1952, for example, Adlai Stevenson, the Democratic candidate, got just 44 percent of the vote against Republican Dwight D. Eisenhower but 64 percent of the Jewish vote; four years later, in their 1956 rematch, Stevenson got 42 percent of the overall vote but 60 percent of the Jewish vote.[59] Jews as a group certainly became less radical, though. Former leftist Harry Gersh, writing in *Commentary* in 1954, confessed that the class enemy, as it turned out, hadn't been what they'd thought. But though many American Jews of the postwar period largely abandoned the politics of class, they took on other liberal causes. One such issue was separation of church and state in schools. In 1957, for example, the Jews of Plainview, New York, objected to a school policy statement that defended Christmas parties and nativity scenes as part of the school curriculum (their sympathetic neighbors sent in letters to the local paper decrying Jews as "atheists" and "godless Communists").[60] Another political cause for many Jews was opposition to the Vietnam War. The Union of Orthodox Jewish Congregations broke with most Jews on this point and voted to endorse the war in 1966, explicitly praising US determination to resist Communist aggression. (Rabbi Abraham Joshua Heschel, a leader in the civil rights movement, was unbowed, and declared, "To speak about God and remain silent on Vietnam is blasphemous.")[61] And the most cherished, and most fraught, liberal Jewish cause was civil rights.

Jews did not lose their liberalism. In hindsight, one wonders if there was not a hint of wishful thinking in Glazer's prediction that they would: that Jews would drop this set of politics that set them apart from other white Americans and join white America's political ranks.

Nathan Glazer argued that the Lower East Side, with its radical, Socialist politics and Yiddish and ethnic Jewish identity, was where Jews were at their most Jewish. He also argued that Jews belonged to white, mainstream America and that they would become more conservative and vote more like other white people. Either Nathan Glazer was arguing that Jews should become less Jewish, or he was arguing that two separate existences, in conflict with each other, were each the true Jewish way. In reality, both were Jewish ways of being, because there were Jews who did both. Sometimes the same Jews did one and then the other. To write about the Jews as Glazer did suggests that there was one true path toward Jewish authenticity. But people travel differently. Jews are no exception. Even if Glazer's visions of how Jews should be were not at odds, they would still be an oversimplification. A people are not a monolith. They can't be pinned down with words like "authenticity."

The ambivalence toward how Jews should exist in relation to class, politics, and socioeconomics more generally manifested itself differently for Jewish men and Jewish women. The American Jewish dream house had different roles—nightmares in their own ways—for men and women to act out. Part of America's war on Communism was touting the wholesome American family, in which women stayed home and raised families for which men went out and provided. American culture and media told women that a ring was more important than a higher degree. This did not apply only to Jews. But for Jewish families, this was as much a shift as the move out to the suburbs.

In the postwar period, married Jewish women actually dropped out of wage work at a slightly higher rate than their non-Jewish counterparts; in 1957, 27.8 percent of married Jewish women worked for wages, compared to 29.6 percent of non-Jewish married women.

For many Jewish married women, that did not mean sitting at home and twiddling their thumbs. Those new synagogues and new congregations that were built had social committees and fundraisers and boards, and women were more than twice as likely as their male counterparts to hold a leadership position in a local Jewish organization.[62] People bemoaned the loss of a Jewish culture, but to the extent that there was Jewish life in the suburbs, it was Jewish women who breathed it into existence. My nana was on her synagogue's ritual committee and an active leader in the synagogue's Sisterhood.[63]

This sense of doing something outside the home was actively used by Jewish organizations to recruit women: Hadassah, the Women's Zionist Organization of America, handed out flyers reading, "Hadassah makes you important."[64]

For sustaining Jewish life, for pursuing activities outside the home, Jewish women were criticized. A 1959 guide for Jewish homemakers insisted that Jewish culture had long been dependent on women being totally devoted to domestic life.[65]

Of course, the idea that Jewish women in the home were central to the continuation of Judaism underpinned arguments against intermarriage in the postwar period. If Jewishness in the postwar period was viewed as religious, not racial or ethnic, then it would be lost forever if people moved into families and households that practiced some other religion.

"It was beaten into us from day one," my father said of how he and his brother were raised. "That we had to marry a Jewish girl."[66]

Intermarriage was increasing, though. Before 1940, the rate of Jewish intermarriage was between 2 and 3.2 percent; between 1941 and 1960, it doubled, to around 6 percent.[67] Some understood which way the winds were blowing. In the story on which part of *Fiddler on the Roof* is based, Tevye's daughter comes back from her interfaith relationship to the family and tradition. In the musical, Tevye, after at first refusing to give his blessing and saying his daughter is dead to him for marrying a gentile, gently wishes her husband and her well: "God be with you."[68]

In 1957, Rabbi Max Eichhorn argued in writing that rabbis who refused to perform interfaith marriages wouldn't get people to see things their way but would rather alienate the couples, who would go ahead with their marriage and carry with them resentment of that rabbi and maybe Judaism generally.[69] But that was not how most rabbis regarded the matter in 1957.

Even those Jewish women who stayed at home to play the part of

the good domestic Jewish wife to their Jewish husbands were criticized. They were materialistic, their henpecked husbands working
to the bone to support their American, suburban way of life. The Reform movement's Albert Vorspan and Eugene Lipman warned that
this materialistic society would make households spend too much
to "keep up with Joneses," which would force women back into the
workforce and out of their homes, compromising the "security" of
their children.[70] American Jewish women in movies and television
weren't depicted as Zionists or activists but as wives and mothers.
The stereotype of the Jewish American Princess was born. "What
does a JAP make for dinner?" the joke went, "A reservation. How do
you give a JAP an orgasm? Scream, 'charge it to Daddy.'"[71]

Later, Nora Ephron would take to task the men who invented
the Jewish American Princess. "You know what a Jewish prince is,
don't you?" she asked in *Heartburn*, a 1980s novel reflecting, in part,
on 1950s Jewish American life. "If you don't, there's an easy way to
recognize one. A simple sentence, 'Where's the butter?' Okay. We all
know where the butter is, don't we? . . . But the Jewish prince doesn't
mean 'Where's the butter?' He means, 'Get me the butter.' . . . I've
always believed that the concept of the Jewish princess was invented
by a Jewish prince who couldn't get his wife to fetch him the butter."

It's a cutting description, and Nora sticks the knife specifically into
Jewish men. But, as Brodkin notes, in the 1950s and '60s women like
Betty Friedan and Wini Breines made their feminist critiques in *The
Feminine Mystique* and *Young, White, and Miserable*, respectively, as
white suburban women, not explicitly as Jews.[72] And so the concept
of the Jewish American Princess was left largely unchecked.

But if Jewish women were flattened into shrewish and materialistic stereotypes, Jewish men were not spared from hand-wringing
over gender identity and socioeconomic status. Middle-class men
generally went through a period of crisis in the postwar period, with
authors wringing their hands over how the comforts of suburban life

had made men soft, but Jewish men went through their own existential crisis.

The concept of the man as the breadwinner in the heteronormative family was a relatively new one in Jewish history and one that clashed, for some, with how the traditional Jewish man was meant to be. There was meant to be more to being a Jewish man than earning money and material things. Philip Roth, who, as much as anyone, made American Jewish themes and writing mainstream in American literature, explored this division in *Goodbye, Columbus*, in which the protagonist, Neil Klugman, ultimately turns down life with his affluent girlfriend, including working for her upper-middle-class father, and returns to his books at the library. The intellectual, like the spiritual (rabbis moaned their lack of status in the postwar period), was seen as lost to the material in postwar America.[73]

Both American Jewish men and women were presented negatively next to their heroic Israeli counterparts. Jewish women were not leading lives of strength and service. Gone, the narrative went, were women like Clara Lemlich, the plucky seventeen-year-old who announced a strike in Yiddish in the garment industry in 1909. In fact, Clara Lemlich didn't disappear. She got married and became Clara Lemlich Shavelson but continued to be a Communist Party member and political organizer, focusing her efforts on working-class wives and mothers.[74] And the American Jewish man was presented negatively next to the tough Israeli soldier. American Jewish men could make money or lead a life of mind, but they could not protect, not really, not like an Israeli could.[75] Never mind that thousands of American Jews had served in World Wars I and II. And to a certain extent, this, too, has stuck in the American imagination. One interviewee, who asked to be identified by only his first name, Max, said that, ahead of our interview, he was thinking about how American Jewish men are emasculated. If you played sports, he said, "there would be—even if you were good—a perception that Jews own the team. They don't play for it."[76]

In the postwar era, as different as their experiences and expectations in this period were, Jewish men and women in America had something in common: both were deemed less authentic than their counterparts in Israel. As American Jews moved out to the suburbs, Zionism moved to the core of their politics.

Chapter 3

Zionist Jews

I N 1906, AN EIGHT-YEAR-OLD girl named Golda moved from Kyiv to Milwaukee. As a child, she had witnessed a pogrom, and, even in America, she did not forget what it felt like to be Jewish, outnumbered, and at the mercy of others. And so, in 1915, she joined the Poale Zion, or Labor Zionists.[1]

In 1921, Golda Mabovitch moved to a kibbutz, or Socialist, Zionist agricultural settlement, in Palestine. Some 24,000 Jews were in Palestine in 1882. That number had increased to 94,000 by 1914. In 1922, the year after Golda arrived, there were 83,794; in 1939, the year World War II broke out, 449,000. By 1947, the Jewish population in Palestine was 630,000, or roughly a third of the total population.[2]

Israel became a state in 1948. Golda remained there. She married. In 1969 she made history when she, now Golda Meir, became the fourth prime minister of Israel.[3]

Golda Meir's story is a remarkable one, and not only because she became prime minister. The very fact that she moved to Israel made her the exception, not the rule. Most American Jews, even those who were fervently Zionist, did not actually relocate to Israel.

When Golda Mabovitch went to America, Zionism was a fledgling movement. But by the time Golda Meir became prime minister, most American Jews would call themselves proud Zionists. The change was in part a reaction to the Holocaust and the tragedy and trauma experienced in Europe and around the world. But not only. The story of how Zionism moved from the fringes to the mainstream among American Jews is also one of assimilation and acculturation.

In the late nineteenth and early twentieth centuries, Jews in America were wary of Zionism. There were, of course, exceptions. But generally, a desire to be white and be seen as belonging in America was the reason that many American Jews, and particularly the more acculturated American Jews, were wary of Zionism in the pre–World War I period. The elite American Jewish Committee started out as staunchly opposed to the Zionist movement,[4] undoubtedly fearing that it would hurt the cause of Jewish assimilation.

The 1914 election of Louis Brandeis to the presidency of the Provisional Executive Committee of American Zionists marked a turning point[5] (he also put up a Christmas tree during the holiday season, making him, according to some family members of people I interviewed for this book, a Bad Jew).[6] On the eve of the First World War, Brandeis tried to make the argument that American patriotism and Zionism were compatible. "Let no American imagine that Zionism is inconsistent with patriotism," Brandeis said. Brandeis stressed that the Zionist movement was a democratic one and that the movement, centered as it was (per Brandeis) around justice and democracy, was thus compatible with America's most cherished ideals. These sentiments were expanded on by Judge Julian Mack, who said, "To be good Americans, we must be better Jews, and to be better Jews, we must become Zionists."[7]

Other high-profile members of the Jewish community also embraced the cause. Rabbi Stephen S. Wise—the Reform Jew who argued against restrictive immigration legislation—was one such someone. Wise was born in Hungary in 1874 but was raised in New York, the son of a Conservative rabbi. But while serving in Portland, Oregon, Wise became ordained as a Reform rabbi and then led a congregation. He was also an early devotee of Zionism. He founded the American Jewish Congress, which oversaw the populist, democratic, and pro-Zionist alternative to the non-Zionist American Jewish Committee. In 1922, Wise opened the Jewish Institute of Religion in central Manhattan. Even after the Institute merged with Hebrew Union College in the 1950s following Wise's death, it influenced a generation of young Reform rabbis.[8]

In 1937, new guiding principles of Reform Judaism were approved in Columbus, Ohio. Whereas classic Reform stressed Judaism as a religion, this new document stressed that it was referring to a "Jewish people," a term that encompassed both ethnicity and faith. Also, unlike the Pittsburgh Platform of 1885, the Columbus treatise referred to a Jewish homeland (though it did not mention political Zionism by name).[9]

Even the American Jewish Committee, composed of elite, self-appointed leaders of the Jewish community who had previously held the belief that American Jews shouldn't publicly support a Jewish state, got on board with the idea, abandoning the notion that one could not be a Zionist and a Good American Jew.[10] Fear that they would be seen as Zionists, not Americans, subsided. American Jewishness and Zionism were beginning to be seen in tandem, not in opposition. The floodgates were opening. In the 1920s, American Zionism as a movement was still relatively small and consumed by schisms.[11] Some had different ideas for what, exactly, Zionism looked like: while some supported a state, others preferred the idea of what would now be called a binational confederation.[12] And Zionism wasn't even the sole Jewish movement for the Jews of Europe. In addition to those in the Bund, a Jewish Socialist movement,[13] there were also Jews who wanted an autonomous region for Jews—but in Europe, not in Palestine.[14]

But then, of course, came the flood.

While American Jews were changing their attitudes toward a Jewish homeland in Palestine, Palestine was changing, too.

On November 2, 1917, Arthur Balfour, the British foreign secretary, wrote perhaps the most famous letter in the history of British colonialism to Lionel Walter Rothschild, of the famous banking family and a leading figure in the Anglo-Jewish world. In the letter, Balfour said that Britain supported "the establishment in Palestine of a national home for the Jewish people." The Balfour Declaration, as it would later be known, did not call for the reconstitution of Palestine as the Jewish homeland, nor did it protect the political rights of non-Jewish communities by name (though it did stipulate, "nothing shall be done which may prejudice the civil and religious rights of existing non-Jewish communities in Palestine"). The British meant to encourage Jewish—and particularly Jewish American—support for the Allied cause in World War I.

At the time, Palestine was part of the Ottoman Empire, against which Britain was fighting in World War I (hence the hope for support by the United States for Britain). The Treaty of Versailles formally awarded Britain the mandate to govern Palestine with the inclusion of the Balfour Declaration.

In 1939, the British changed their policy toward Palestine, recommending no more than 75,000 more immigrants to the region and an end to all immigration by 1944 barring the support of Palestinian Arabs. Zionists considered this to be grossly favoring the region's Arab population.[15]

Then the Holocaust happened. Six million of Europe's Jews—

those who had not been allowed to come to the United States and did not go to Palestine, those who were forced to stay in Europe, and those who chose to stay because they thought that Europe still held the answers for them—were murdered.

"I remember saying to my father, 'You only vote for Israel. Why aren't you voting for what's best for the United States?'" my father said when he told me of debates he had growing up with his father, Alvin Tamkin, Solomon and Fannie's son. And he came to believe that Israel was his father's main political concern because "they all grew up during the Holocaust. American Jews had enormous guilt. They were all over here being fat and happy while the others were being killed and we didn't do anything about it. Morgenthau didn't do anything about it. . . . All these children of the Diaspora during the Holocaust," my father said, "I think were really, really committed to Israel. Because of guilt."[16]

Where my nana's response to the trauma of World War II was to throw herself into making sure her children received a Jewish education and married Jews, my grandfather made Israel the center of his political life. Many in his generation were wracked by guilt that they and the United States and the world hadn't done enough to support Jews before and during the war. They told themselves that they would not let history repeat itself. This would never happen again. Not to them, anyway. Not to Jews.

There was no one to advocate for Jews in the Truman White House. Roosevelt had Morgenthau (whose efforts to help Europe's Jews during World War II were both insufficient and thwarted), but he was gone. That changed when a B'nai B'rith official realized that a member of the organization in Kansas City was an old friend of the president. Enter Eddie Jacobson.

Jacobson was no Zionist, but said that he would do what he could. And he did. He got Truman to meet with Chaim Weizmann, a Zionist leader who would become the first president of Israel ("You win, you bald-headed son of a bitch," Truman told Jacobson after relenting to the meeting). Warren Austin, the US ambassador to the United Nations—without White House approval—recommended abandoning the plan to partition Palestine. But then, on May 14, 1948, eleven minutes after David Ben-Gurion proclaimed the birth of a Jewish state, the United States officially recognized Israel. Eddie Jacobson sent a message to Truman: *Thanks, and God bless you.*[17] (Some American Jews, known as the Sonneborn Institute, had previously taken matters into their own hands, committing funding to purchase arms to equip the new Jewish army.)[18]

For Jacobson, the moment was one of joy. For hundreds of thousands of Palestinians, great pain.

In May 1948, Israel's founders promised, in their "Declaration of the Establishment of the State of Israel," "complete equality of social and political rights to all its inhabitants irrespective of religion, race or sex." Theodor Herzl, the father of modern Zionism and the idea of the state of Israel (though he died before its establishment), wrote, "We don't want a Boer state, but a Venice," meaning not a state divided by racist nationalism, as the Boer states were, but a place of tolerance, even enlightenment. The lofty ideals were one thing; the reality on the ground in the wars that preceded and followed the establishment of the state of Israel was another. David Ben-Gurion, Israel's first prime minister, was "shocked by the deeds that have reached my ears."[19]

Seven hundred thousand Arabs left Palestine. While some may have left of their own volition, not wanting to live in this new state, others were forced out. According to Tom Segev's *1949: The First Israelis*, in the town of Jish, in Upper Galilee, soldiers pillaged Arab houses. Those forces brought those who protested to a remote location, where they shot and left the protesters dead.[20]

For many Jews, the establishment of Israel was a miracle, a relief, a cause for celebration. But in Arabic, the exodus of seven hundred thousand Arabs from Palestine is called the Nakba, or catastrophe. Ben-Gurion may have claimed to have been horrified, but he also ordered Yitzhak Rabin, then a military commander and later prime minister, to drive out fifty thousand civilians from Lydda and Ramla in two days in June, per Rabin's memoir.[21]

In 2019, Rashida Tlaib, a newly elected Palestinian-American member of Congress, drew ire for saying on a podcast that she had a "calming feeling" when she stopped to consider that her ancestors had lost their land and livelihoods because "[a]ll of it was in the name of trying to create a safe haven for Jews, post the Holocaust, post the tragedy and horrific persecution of Jews across the world at that time. And I love the fact that it was my ancestors that provided that, right?" Some lambasted her as an antisemite, while others said she was being ahistorical. Aaron David Miller, who had advised presidents from both parties on the Middle East, said, "Every institution for what would become the state of Israel was in place well before Hitler started killing any Jews."[22]

Tlaib clarified what she meant on a late-night talk show interview with host Seth Meyers. Her grandmother lives in the West Bank, she said. "The tragedy of the Holocaust—the reason why Israel was created was to create a safe haven for Jews around the world. And there is something, in many ways, beautiful about, that my ancestors—many had died, or had to give up their livelihood, their human dignity, to provide a safe haven for Jews in our world. And that is something that I wanted to recognize and kind of honor in some sort of way." She added, "I want Palestinian people also to find some sort of light in what's happening." And listening to her, I didn't hear antisemitism or hatred for Jews at all. I heard a person who was trying to make sense of loss and pain.

"In the end," she said, "I want all of us to feel safe. All of us deserve human dignity."[23]

. . .

Israel became a new state. Ben-Gurion and his government built state institutions and helped the new country take in Jews from around the world.

Arabs who managed to stay made up less than 20 percent of the population of the new country. They were given Israeli citizenship but lived under martial law for almost twenty years.

In the first two decades after the establishment of the state of Israel, American Jewish support was, for the most part, muted. American Jews, after all, supported the existence of a Jewish state but didn't see themselves as Jewish refugees in need of a homeland. They had a home. It was the United States.

"My feeling was always very strong for [Israel]," my high school mentor Arlette Sanders told me. "It was very exciting to my family, and to me as well. The knowledge that there was a homeland for Jews that they could all go back to if they wanted.

"I was very happy it was there," she said. "But I identified with America. . . . To me it's important there is a homeland for Jews, I had very strong feelings about that, but not a place I wanted to go to [live]."[24]

In contrast to Golda Meir, Arlette Sanders's impulse typified the view of American Jews. There were Americans who did move to Israel, and who believed that doing so was the only way to truly live a Jewish life (as Hillel Halkin wrote in his 1977 work, *Letters to an American Jewish Friend*).[25] But more common was the American Jew who, like Sanders, liked the idea of having a Jewish state, or who thought it was important for a Jewish state to exist for Jews who survived the Holocaust, but they personally did not want to leave the United States. They considered themselves not a Diaspora in need of a homeland, but Americans who were Jewish.

That wasn't to say support for Israel didn't exist: it did. Some synagogues introduced prayers for Israel and started flying the Israeli flag.

But through the 1950s, Israel was still in American Jewry's periphery. A study of Jewish education toward the end of that decade found that only forty-eight of one thousand Jewish educators reported teaching about Israel as a subject. A 1951 survey lamented, "Not a single book has been published since the establishment of the Jewish State which deals with any of the numerous social aspects of the great events taking place at present in the country."[26] Israel did have some influence on Jewish institutional and philanthropic life. One notable example was Hadassah, a Zionist women's organization. But it was successful not only because of its pro-Israel position but because its organizers understood the needs of American Jews; Hadassah provided a Jewish context for women's social lives.[27]

American Jewishness and Israeli Jewishness did not always coexist. Sometimes, they collided with each other.

In 1960, the Mossad, Israel's intelligence agency, captured Adolf Eichmann, one of the masterminds of the Holocaust, in Argentina. Israel, naturally, wanted to try to convict him in Israel.

American Jews had other ideas. The American Jewish Committee had already made the switch from being anti-Zionist to pro-Israel, but its former president, Joseph Proskauer, nevertheless urged David Ben-Gurion, the first Israeli prime minister, to hand over Eichmann to an international tribunal. Proskauer's reasons were twofold. First, he did not believe that Israel should speak for or in the name of the world's Jews. And second, Eichmann committed crimes not just against Jews but against humanity. Similarly, Rabbi Elmer Berger, leader of the anti-Zionist American Council for Judaism, didn't want Israel to try Eichmann because he knew it would make the country the center of Jewish gravity. Jewish life and Jewishness and Jewish identity—in Berger's mind, this didn't belong to Israel, but to Jews all over the world, including in the United States.

Ben-Gurion was evidently disgusted. "Israel," he shot back, "is the only inheritor of those Jews [murdered in the Holocaust] for two reasons; first, it is the only Jewish state. Second, if these Jews were alive, they would be here because most, if not all of them, wanted to come to live in a Jewish state."

The second statement was not, in fact, true. Europe's Jews had not all been clamoring to pick up and move to Palestine, and many saw themselves as either assimilated or Socialist or preferred to agitate for a Jewish homeland while residing in Europe. But Ben-Gurion's belief that Europe's Jews inevitably would have wanted to come to Israel spoke to his frustration with the reality that America's Jews were not moving to the only Jewish state.[28] Ben-Gurion came to resent many American Jews; even those who called themselves Zionists, he realized, did not actually intend to leave their comfortable American lives to fulfill the promise of Israel.[29] Zionism, for most American Jews, was a political belief, and a dream, and something they would work ardently at home to support, but it would not be a lived experience. American Jews wanted a Jewish homeland, but most did not want to make that land their home. American Jews would compare themselves unfavorably to Israeli Jews—Israeli men were tougher, Israeli women were more useful—but most would not actually move to Israel.

And even though most American Jews did support Israel, they disagreed over how Jewishness should manifest itself in Israel.

Traditional Orthodox groups, like the Satmar Hasidim, who established themselves in Brooklyn after World War II, were anti-Zionist.[30] Traditionally Orthodox Jews also opposed specific Israeli practices, picketing the Israeli consulate in New York; they were opposed to the Israeli army's drafting of women.

Orthodox Jewish groups in the United States opposed efforts by Conservative and Reform Jews to establish themselves in Israel and defended those Israelis whose religious interpretations were questioned by more liberal denominations.[31] (It was through his support of Israel, Arlette Sanders told me, that her father became involved with "very Orthodox people. The next thing I knew, I was not allowed to go out on Friday night. For show. Completely for show.")[32]

Israel became one more way in which the differences between Jews played out. And it was not only that Jews of different sects had different opinions on what should or should not be happening in Israel. It was that there was a disagreement over who had a right to be invested in Israel, and who got to speak about Israel, about which of America's Jews had the right to have an opinion on the Jewish state.

In the early 1960s, American Jews became even more aware and conscious of Israel. This did not particularly impress Ben-Gurion, who said in 1960, "Judaism of the Jews of the United States and similar countries is losing all meaning, and only a blind man can fail to see the danger of extinction, which is spreading without being noticed."[33] American Jews were by and large undeterred by such scorn. The 1960 film *Exodus* made Zionist pioneers seem like the heroes of American westerns and presented the country as a kind of democratic oasis. And there was a concerted effort by the United Jewish Appeal, the central communal philanthropic Jewish organization, to train young Jewish leaders to take pride in Israel.[34]

Still, the kind of consensus on Israel that came to characterize large swaths of Jewish American life had yet to form. And then came 1967.

Tension was high on the Syrian border in the spring of 1967. The year so far had been marked by the shelling of kibbutzim. On May 13, the Soviet Union warned Cairo that Israel would attack within a week.

Egypt began remilitarizing the Sinai Peninsula. It closed off the Straits of Tiran to ships coming in and going out of Israel. The UN withdrew its forces. France and Great Britain declared that they would no longer stand by the 1957 agreement declaring free passage.

On May 30, Jordan pledged troop support to Egypt, and troops from Iraq, Saudi Arabia, Algeria, Kuwait, and Egypt all gathered in Jordan close to that country's border with Israel.

And then, on June 5, the Israeli air force attacked the other forces' airfields. Egypt, Jordan, Iraq, Saudi Arabia, Algeria, and Kuwait all lined up against Israel, and it took six days for Israel to win.[35]

The experience of the Six-Day War united and galvanized many of America's Jews. What if the war hadn't been won? What if Israel had been destroyed?

Day-to-day life would not have changed for most of America's Jews, most of whom, again, had no plans to move to Israel, although some now had family members and friends there. But the Jewish state that so many—like Arlette Sanders—were just glad to have *there* would be gone. Jews would once again be crushed or killed or uprooted or some combination thereof. And American Jews would again be left with their comfortable lives and their guilt. And so many of them set about trying to do what they felt they hadn't a few decades prior: they tried to do something.

The United Jewish Appeal raised $100 million in one month for Israel. At a New York luncheon, five people pledged a total of $5.5 million in fifteen minutes. Israel had become *the* political cause of American Jews.[36]

Concerns for Israel's safety offered American Jews a something around which to come together—and a new narrative about themselves. It was as though, in the 1960s and '70s, Israeli Jews created a new model for how Jews could be perceived: strong and tough (never mind that thousands of American Jews had served in the First and Second World Wars). The 1971 movie version of the Broadway musical *Fiddler on the Roof* recast the starring role, Tevye. The director, Norman Jewison, cast Topol, an Israeli actor. In the documentary on the musical, *Fiddler: A Miracle of Miracles*, Paul Michael Glaser, who played Perchik in the film, says that every previous Tevye had played the role in what he, Glaser, described as the "Eastern European Jewish tradition." When he says, "Why?" it's with a whiny shrug. Topol, by contrast, played it the Israeli way, per Glaser. "'Why!' He'd demand an answer," Glaser said.[37]

That Eastern European Jews were weak and whiny was a stereotype, one that didn't take into account, for example, the Warsaw Ghetto uprising of 1943 (distinct from the Warsaw Uprising of 1944), in which Jews in Nazi-occupied Poland, knowing they would lose, tried to fight back against the Nazis.[38] But the idea of the strong Israeli Jew and the weak Eastern European Jew (and, by extension, the weak American Jew whose ancestors came over from Eastern Europe) nevertheless stuck.

The perceived heroism of Israel—particularly after the Six-Day War of 1967—led some Jews to identify more openly as Jews. It also moved Israel to the center of their Jewishness. Hebrew schools, Jewish community centers, and Jewish social service agencies all started decorating their buildings and offices and classrooms with posters of Israel. Most American Jews started using and teaching Israeli-style Hebrew, which was more influenced by Sephardic Hebrew, as opposed to Ashkenazic Hebrew. They incorporated Israel into formal programming: summer camps, for example, began hiring Israeli camp counselors.[39] The Jewish Publication Society published twenty-eight Israel-related books in the decade following the Six-Day War, more than it had in the previous two decades combined.[40]

This trend was only exacerbated by the 1973 War. An Egypt-led Arab coalition launched a joint attack on Israel on Yom Kippur, which, in the Jewish calendar, is a day of fasting, prayer, and reflection.

"That was during Yom Kippur," my father told me, adding, "we were in temple, obviously, and everyone, there were radios playing, which was unheard of, in temple, monitoring the battle that day. . . ."[41] (In *We Stand Divided*, his book on American and Israeli Jews, Daniel

Gordis also recounts a radio in the synagogue on the day the Yom Kippur War launched, suggesting that, though without precedent, it was also not exactly a rare event on that day.)[42]

American Jews and American Jewish organizations raised $107 million in the first week of that war. Over the course of the war, which took less than one month, they raised $675 million.[43]

In 1944, the National Community Relations Advisory Council, the umbrella organization of most Jewish groups, left "Jewish" out of its name. In 1971, they added "Jewish" in, becoming the National Jewish Community Relations Advisory Council.[44] My grandfather went on a B'nai B'rith tour in 1967 shortly after the war and came home with a photo of himself in a phantom jet. My father remembers asking him in 1968 and again in 1972 whether the only political issue he, my grandfather, cared about was Israel.[45]

I am far from the only Jewish person of my generation to have such a story about their grandparents. The journalist Abraham Riesman wrote of his grandfather Robert Arnold Riesman Sr., a Jewish philanthropist and staunch Zionist, "In his eyes, Israel was always under mortal threat, and if its foes were to defeat it, there would be mass Jewish death there on a scale with which his generation was all too familiar. If his people lost their citadel in the Middle East, who knew what other dominos might fall? The scion of his own line could be next."[46]

None of this is to say that tensions, complications, and profound differences didn't remain between American Jews; they did. So, too, did tensions exist between American and Israeli Jews. In May 1972, my father and his parents took a trip to Israel. "I went right before my bar mitzvah. We went on a three-week trip, which was quite an incredible trip," he told me. He thought for a beat, and then added, "That was actually kind of a shocking trip."

It was shocking both because of the materialism exhibited by some American Jews on the tour (my father's own observation it-self an example of American Jews' discomfort with this Jewish ste-reotype) and also because of the attitude exhibited by Israeli Jews toward Americans. "The other thing I got on that trip was how much the Israelis—the *sabras*," he said, using the word for Israelis born in Israel, "the first-generation or second-generation children of the Ho-locaust survivors—they were tough as hell, they had fought in wars, they had been in the Israeli army in '48, '56, '67. They really resented American Jews.

"Even though American Jewish money was helping pay for the whole place, they really resented them. They hated them." Israeli Jews were the ones who were there, the ones getting attacked, the ones living in the Jewish state American Jews only talked about. One Israeli asked my father what American Jews would do if Mexicans sent rockets across the border.

His tour guide had been a frogman in the Israeli military, placing explosives around the harbor at Alexandria during the Six-Day War, or so he said. The bus driver had been a tank commander. They made fun of everyone on the trip and "were horrified when I spoke Hebrew to them," my father told me, since they realized that my father could understand what they were saying about their American clients. Af-ter that, he and his parents got to sit in the good seats on the bus in exchange for my father not sharing their insults with the rest of the group. (My nana got into a confrontation of her own with the tour guide, who was taking the group shopping instead of to sites like the birthplace of Jesus. Jews didn't want to see that, he told her. At which point my nana, a "Conservadox" Jew, demanded to be taken to Beth-lehem, Nazareth, and East Jerusalem.)

They went to a kibbutz, the collective, agrarian societies in Israel that were, at first, a blend of Socialism and Zionism (they were later privatized). Their waiter, he told me, was an Arab man. When asked why he was the one serving food at this Jewish Socialist endeavor, the waiter said he was there because the Israeli Jews didn't want to serve American Jews.[47]

My father was not the only one to perceive a distinction between American and Israeli Jews. Jo-Ann Mort is a journalist and author turned communications consultant based in New York. Her religion, she told me, is Israel. She has so many clients and friends and family there that, she says, "My day is split between there and here, no matter where I am."

She was struck, she said, on her first trip there. "I went there and I had the reaction, 'Oh my God, everybody's Jewish!'" Everybody isn't, she quickly clarified; now, she said, over half of her friends over there are Palestinian. Still, "I came back so in love with Israel, so comfortable there. I had lunch with Irving Howe."

She and the Socialist writer met at Leo's on Eighty-fourth and Madison in New York. "How can I not live in Israel?" she asked him.

"He said, 'Jo-Ann, you cannot live there . . . they're not people like us.'"

She was not, he was telling her, going to fit in in Israel. She was Jewish, and Israelis were Jewish, but they were not the same kind of Jewish people.

"'You'll have what Jews have had for thousands of centuries: a sense of homelessness,'" she recalled him telling her.

Mort doesn't think that was actually true for Howe. He wasn't homeless: he had Yiddishkeit, had the Yiddish language and culture and Socialism.

Even so, she came to believe that what she was looking for—belonging, maybe, or authenticity—"I couldn't find that in Israel." Israel became a passion and a cause for American Jews, but, for most of them, it did not give a sense of home or even of a new or right or better way to be Jewish.

"So I came back here," she said. But she still feels like she lives in between the two worlds, with half of her there, in Israel, where she travels often, spending weeks at a time. She is an American Jew, she told me, but one who thinks about Israel every day.[48]

Even with these complications and complexities between the two identities and places, whether one was a Good or Bad Jew was coming to depend not only on one's Americanness but also on one's connection to a state a world away. As Riesman's grandfather said of the country, in the early years of Israel's sovereignty, "It was really like a relative that you had to support, whose company you didn't particularly enjoy, who gave you no excitement, no stimulation." As the twentieth century passed, though, he came to believe, "It was because Israel was threatened that it became precious. . . . When it wasn't threatened, it was an inconvenient relative; when it was threatened, it became something you liked."[49]

This phenomenon happened in concert with another. Zionism would change many American Jews' relationship to the civil rights movement—and the civil rights movement would, in turn, change its relationship with Jewish identity.

Chapter 4

Civil Rights Jews

I T IS PERHAPS THE most famous black-and-white photograph in American Jewish history: Rabbi Abraham Joshua Heschel walking alongside Martin Luther King Jr. Heschel marched from Selma to Montgomery, Alabama, in 1965. When, on his return, he was asked if he found time to pray, he replied that he prayed with his feet.[1] Hundreds reportedly wore yarmulkes out of solidarity with the rabbis marching that day. Five rabbis were put in jail for the march; from their cells, they recited Hebrew prayers.[2]

In the narrative of American Jewish postwar history, few episodes hold a place of as much pride or prominence as Jewish support for the civil rights movement. When Reverend Raphael Warnock, who is Black, and Jon Ossoff, who is Jewish, were both elected to the US Senate from the state of Georgia in January 2021, Warnock said that he believed Heschel and King were smiling down on them.[3] The memory of that day, and of that moment in Black, Jewish, and American history, still lingers.

The photograph endures in part because it encapsulates one of the most enduring stories of American Jewish history: That American Jews understood the importance of civil rights and staunchly supported the movement in the 1950s and '60s, or at least until Zionism and Black nationalism drove Jewish and Black Americans apart. That American Jews viewed achieving civil rights for all, including Black Americans, not only as a cause worth fighting for in and of itself, but as a necessary part of securing a free society for themselves. "It is not always the same race, color, or creed that is subjected to abuse," said Isaac Toubin, associate director of the American Jewish Congress in 1953, "but this abuse, no matter what its target, always poses the identical threat to the achievement of a peaceful and just communal life."[4]

This is one story. But it is not the whole or only story. Some American Jews opposed civil rights or sat silently by. Some supported civil rights in word but not in deed. And some supported the movement— and, in doing so, risked ostracization from the mainstream American Jewish community.

It is true that many American Jews in the 1950s and '60s said that support for civil rights was a mark of being a Good American Jew. Polling data from the late 1950s suggests American Jews considered support of civil rights more important to being a "Good Jew" than Zionism.[5] Support for civil rights also helped back a vision that many American Jews had of themselves: a minority that stood with other minorities, a persecuted people who helped other persecuted people and also, critically, who were American enough to take part in an American debate as to how to be a country.

American Jews made up roughly 3 percent of the population and constituted two thirds of white Freedom Riders (activists who rode interstate buses to protest the segregation of public transport in the segregated South) in 1961 and one third of the white student volunteers who went down to Mississippi to register voters in 1964. That summer, two Jewish men, Andrew Goodman and Michael Schwerner, were murdered along with James Chaney, a young Black man.[6] Schwerner and Chaney were organizing as part of the Congress of Racial Equality, while Goodman was volunteering to help further voter registration and civil rights.[7] The investigation of their deaths inspired the 1988 film *Mississippi Burning*, and the summer arguably spurred the passage of major civil rights legislation.[8]

Paul Cowan, a recent graduate of Harvard University in the 1960s, said that he participated in the civil rights movement because of what he called his mother's "secular messianism: a deep commitment to the belief that we had a lifelong debt to the six million dead." It was this debt, he later recalled, that led him to participate in voter registration drives during Freedom Summer.[9]

Speaking at a conference on "The Challenge of Jewish Youth: Israel and America, 1965," Cowan argued, "a commitment to the civil rights movement is a greater moral duty than coming to Israel merely because one is a Jew." What's more, he said, it was through this activism that he was reaffirming his own Jewish identity. *"That,"* he said, "is my own way of resisting assimilation." (This was rejected by some in attendance. Jane Satlow Gerber, a panelist at the conference, said, "The burning issue for Jews in America is to find some kind of relevance within their tradition. This is a more imminent problem than that of a few thousand Jews in the civil rights movement.")[10]

More high-profile American Jews participated in the civil rights movement, too. For instance, Bella Abzug, an American lawyer and, later, a member of Congress, tried in 1950 to defend Willie McGee, a Black man falsely accused of raping a white woman.[11] She was unsuccessful, and he was executed in 1951. And, most famously, there was Heschel's friendship and partnership with King.

"The civil rights movement was also an ecumenical movement," Susannah Heschel, Rabbi Heschel's daughter, told me. It was remarkable for her father, who spent the 1930s in Germany, where political forces tried to disavow the Old Testament and assert that Jesus was an Aryan, to hear King make Exodus and the prophets central themes of the civil rights movement. And the prophets, and Prophetic Judaism—ethical and universalist and full of searching and guidance—were central to Heschel, too.

"What people admire also is that he marched on behalf of people who were not his people. And he made them his people. And he did so in a very strong voice," she said. "That passion that comes through in that speech is the passion of the prophets.

"My father," she said, "gave us a Judaism that's full of passion and emotion." That passion and emotion were tied up with the fight for civil rights.[12]

American Jews got involved in the civil rights movement on both an individual and an organizational level. The American Jewish Congress in particular worked closely with the NAACP (so closely that the two shared reports in 1947) and invested so heavily that it had more civil rights lawyers on its payroll than the Justice Department did.[13] Israel Goldstein, a Conservative rabbi who became president of the American Jewish Congress, declared in 1956, "We must defend the rights of the Negro as zealously as we would defend our rights as Jews whenever and wherever these might be threatened. This is a moral issue from which we dare not avert our faces because of reasons of expediency. It is an American problem in the solution of which all Americans should feel that they have a common stake."[14] American Jews, in other words, were supporting civil rights not despite the fact that they were American but *because* they understood themselves to be Americans, and that these problems were problems for Americans to tackle.

American Jews, at least those who supported civil rights, recognized that, as they were now a part of America, they had a say in how the country should be. There was a more self-interested component to their involvement, too: a pluralistic society, these American Jews appeared to recognize, was better for all people, or certainly anyone who might identify in some way as a minority. Our rights, in a democracy, are really only as protected as those of the least protected person.

But support for civil rights in principle ran into conflict in practice. Most American Jews who lived in the South did not feel secure enough to stand up publicly against racial segregation. Liza Kaufman Hogan, who now lives in Washington, DC, but spent her childhood in the 1960s and '70s in Alabama, recalls a division between a strong strain of liberal politics—like that of her grandparents, who strongly supported the labor and civil rights movements—and those who felt that liberal politics were bad for business, and who just wanted to, as Hogan put it to me, "go along to get along."[15]

They were right that their support for civil rights could interrupt their ability to do that. Norton Melaver, a Jew in Savannah, along with his wife, signed a petition in support of busing students in to integrate the schools. In response, white mothers picketed the supermarkets the Melavers ran. Melaver even recalled that they were threatened: if they didn't give up, something might happen to their children.[16]

Many Southern Jews pressed Jewish organizations to take a moderate approach to anti-segregation activism so that it would not be seen as an especially "Jewish" cause. They also charged that Northern Jews were hypocrites. American Jews in the North, after all, supported civil rights—but contentedly sent their children to segregated public schools. What's more, Jews moving out of the cities into suburbs where Black Americans could not also live contributed to segregation. Many of the large establishment Jewish organizations—namely, the Anti-Defamation League and the American Jewish Committee—agreed with this more muted approach. These were the same groups, as Michael E. Staub wrote in Torn at the Roots, that pushed American Jews to cooperate with the government's anti-Communist activities, and took a "shah shah," or hush hush, approach to dealing with antisemitism. Post-1954, they favored that same "shah shah" approach—staying publicly quiet and ostensibly working quietly behind the scenes—on civil rights.[17]

And while most American Jews supported Lyndon B. Johnson and his "Great Society," some Orthodox Jews, like Rabbi Jerry Hochbaum, questioned whether "racial and urban unrest" were matters of Jewish concern.[18] Nathan Glazer expressed his discomfort with "color-conscious society" that grew from "Black thinking."[19] He also defended Jewish exclusionary practices, saying that they were not racist but rather "part of the standard Jewish ethnocentrism."[20] Glazer—who, as we saw in chapter two, implicitly made the case that Jews should be considered white by comparing Jewish and Black Americans—argued that Jews made their way up in American society "through the principle of measurable individual merit" and that "it is clear that one cannot say the same" about Black Americans. At the 1964 annual meeting of the National Conference of Jewish Communal Service, Glazer argued that the "Negro revolution" posed a threat to Jewish security in America.[21] Morris Grumer, executive director of Jewish Vocational Service in Los Angeles, said that American Jews should "establish safeguards to ensure that in making a contribution toward the Negro community, we are not watering down our service to the Jewish community, and that we are not acting in a manner which begins to destroy our identity as part of the Jewish community."[22]

Relatedly, while many American Jews supported the concept of affirmative action, Jewish institutions had hang-ups about quotas, pointing to their own early-twentieth-century history in which quotas were used to keep Jews out of white elite institutions. Some in the Jewish community tried to make a moral argument—and, indeed, an argument grounded in Judaism—to quash those concerns. A 1967

Yom Kippur sermon by Conservative rabbi Dov Peretz Elkins applied the concept of teshuvah, or repentance, to affirmative action.[23] And in 1963, American Jewish Congress leader Leo Pfeffer argued that American Jews had "no moral right to condemn the Negro for demanding quotas" and that American Jews would demand the same if the roles were reversed. Still, the American Jewish Congress, like the American Jewish Committee, Anti-Defamation League, and National Jewish Community Relations Advisory Council, opposed quotas.[24]

American Jews, now, today, tell the story of our support for the civil rights movement. But the whole picture is more complicated, and more blemished, than that. There was not wholehearted support from everyone in the community nor even the most prominent institutions. American Jews felt proud that Jews were involved in civil rights, Susannah Heschel told me in that 2021 phone call, and "feel proud that my father marched with Dr. King." But to what extent was this pride earned? "What did they *do*? Did they not live in white gated communities, [did they] stick their neck out for affirmative action? No."[25]

In addition to not considering the full picture of affirmative action for Black Americans, American Jews who blasted it also failed to consider how it affected all American Jews—that is, how affirmative action tended to help Jewish women.

Once, Heschel told me, a Jewish college president came to give a talk at her university. He spoke about how bad affirmative action was for Jews. Susannah Heschel raised her hand. She was called on.

"I said, 'I'm a Jew and I'm a woman and affirmative action has helped my career enormously.' And he said, 'Oh, I didn't think of that.'

"Jewish women don't exist in that mentality," she said. "In the mentality of Jewish organizations." Even as those organizations claimed that they were speaking for Jewish people, asserting the Jewish position on behalf of Jewish men and women.[26]

...

Even insofar as American Jews were involved in the civil rights movement, some Black Americans bristled at what they saw as American
Jews' patronizing tone and treatment of Black Americans.

In 1961, the American Jewish Committee invited Whitney M.
Young Jr., who was the new national head of the Urban League, to
speak. "Face the fact that your community does discriminate," he said,
adding, "Words like gradualism and moderation are meaningless."[27]

Others resented the comparison that American Jews tried to draw
between themselves and Black Americans. Jewish history—including
twentieth-century history—is indeed full of persecution and murder
and trauma and hardship. But one's grandfather coming to America
to escape persecution is not the same as one's grandfather coming
to America enslaved. Jews themselves were not wholly unaware of
this: Justine Wise Polier, Rabbi Wise's daughter, warned in 1960 of
intergroup discord and challenged assumptions that Black and Jewish Americans had a similar experience. Five months later, Nathan
Edelstein, governing council chair of the American Jewish Congress,
spoke of deteriorating Black-Jewish relations.[28]

Several years later, in 1966, Bayard Rustin, a Black American civil
rights champion, stood before the American Jewish Congress and
said, "Dear people, when Jewish people run about boasting about
how we Jews made it because we were intellectual, and lifted ourselves by our bootstraps, and we have such extraordinarily beautiful family life that obviously we just went up to the top like cream
in coffee—well, this is hot air." Most American Jews, legally, had always been white in this country. When that whiteness seemed under
threat, American Jewish leaders made the case for their own whiteness. Black Americans could not do this.

"If you are going to remain Jews only so long as Negroes remain
nice," Rustin added, "give it up."[29]

Other Black Americans put forth similar criticisms. "One does not wish, in short, to be told by an American Jew that his suffering is as great as the American Negro's suffering," James Baldwin wrote in his 1967 essay, "Negroes Are Antisemitic Because They Are Anti-White." "It isn't, and one knows that it isn't from the very tone in which he assures you that it is."

Baldwin clarified how Black American and (white) American Jewish experiences meaningfully differed.

"The Jew's suffering is recognized as part of the moral history of the world and the Jew is recognized as a contributor [to] the world's history: this is not true for the blacks. Jewish history, whether or not one can say it is honored, is certainly known: the black history has been blasted, maligned and despised," Baldwin wrote. "The Jew is a white man, and when white men rise up against oppression, they are heroes: when black men rise, they have reverted to their native savagery."

The essay is a powerful one, which notes that "many Jews despise Negroes, even as their Aryan brothers do" and that "many Jews use, shamelessly, the slaughter of the 6,000,000 by the Third Reich as proof that they cannot be bigots—or in the hope of not being held responsible for their bigotry." But ultimately, Baldwin concedes that Black Americans' real issue is not with Jews but with the white Christian world, and so with the part that Jews play in it ("The crisis taking place in the world, and in the minds and hearts of black men everywhere, is not produced by the star of David, but by the old, rugged Roman cross on which Christendom's most celebrated Jew was murdered. And not by Jews.")[30]

But the essay also implies that all Jews are white.

Jews in the United States were, essentially since the advent of the country, coded as white. This was not, however, true for Jews elsewhere in the Americas—for example, in Suriname—and it was not true for Jews around the world more broadly. But even in the United States, Jews were not just white. Black Jews have a long history in the United States. Rufus L. Perry, son of a prominent Baptist minister, converted to Judaism in 1912. In the mid-twentieth century, there was, for example, a congregation of Ethiopian Hebrews in Harlem: the Commandment Keepers, led by Rabbi Wentworth Matthew, that focused, according to Princeton's Judith Weisenfeld, on spiritual liberation and felt Judaism affirmed their humanity in a way Christianity could not.[31] (Americans who identified as Ethiopian Hebrews were distinct from Ethiopian Jews, both of which are sometimes confused with, but are very much distinct from, Black Hebrew Israelites, who believe they are descended from a lost tribe of Israel. Some Black Hebrew Israelites argue that they are the only true Jews.)[32]

And in the 1960s, Jewish racial diversity began to increase further.

Since the restrictive immigration law in 1924, and particularly in the post–World War II period, with a surer footing in the United States, American Jewish leaders had been advocating to have immigration reform enacted such that immigration law no longer effectively discriminated against Eastern European Jews. They found an unlikely champion in John F. Kennedy. Though the Kennedy family patriarch, Joseph, at first insisted that the family be all-American and not play up their Irish roots, the political clan soon realized that "ethnic" Americans, who associated strongly with their own communities' immigrant stories, were also voters. American Jews were one such constituency.[33] In my parents' apartment today, there is a photo of my grandpa, who was, at various points in his career, a state representative, a judge, and chairman of Boston's Jewish Community Council's legislative committee, shaking hands with JFK, on whose campaigns he actively participated.

Kennedy took up the platform of immigration reform, though with no real means to achieve it. When he was assassinated, his successor, Lyndon B. Johnson, understood that the way to achieve

political immortality for both Kennedy and himself was to imple-
ment the platform Kennedy had only talked about. And since Johnson, unlike Kennedy, had spent years in Congress figuring out which
political levers to pull to get legislation through, he managed to pass
immigration reform.[34]

But advocates from the American Jewish community were primarily using contemporary political legislation, with real-world effects, to right a historical wrong, not to improve immigration policy
for the future. As historian Mae Ngai writes in *Impossible Subjects*, "the
new law was not inclusionary towards all. By extending the system
of formal equality in admissions to all countries, the new law affected
immigration from the Third World differently—creating greater opportunities for migration from Asia and Africa but severely restricting it from Mexico, the Caribbean, and Latin America."[35] Many of
the Jews involved in pushing for immigration reform as historical re-
dress had likely not considered that its passage would mean that Jews
from all of these regions would come to America, too, changing the
makeup of American Jewishness.

Two years later, in 1967, the Supreme Court case *Loving v. Virginia*
ruled that anti-miscegenation laws, or laws that banned interracial
marriage, violated the Equal Protection and Due Process Clauses of
the Fourteenth Amendment. Interracial intermarriage was still relatively rare through the 1960s,[36] but that, too, would change in ways
that were perhaps unforeseen by many American Jews in the 1960s.

Still, while most American Jews were then, and are now, considered white under US law, not all are. It's true that, in the mid-century
civil rights era, the vast majority of American Jews, and certainly
American Jews playing a prominent role in the civil rights movement,
were white. And white Jewish Americans and Black Americans in the
civil rights movement tended to speak of each other as two separate
groups. But as time went on and American Jewishness became more
racially diverse, the implication that these were thought of as two

different and separate groups would have consequences for many American Jews, as we will see in later chapters. Consider, for example, that by the time the 1990 National Jewish Population Survey was completed, researchers found 2.4 percent of the sample self-identified as Black; Black Jews themselves have estimated that there are as many as 1.2 million Black Jews in the United States, who necessarily have a different relationship to civil rights history than white Jews do.

"Many white Jews are holding tight to that memory of the civil rights movement," Rabbi Sandra Lawson, a director at the Reconstructionist movement and one of the world's first queer Black rabbis, told me. "You can't keep holding up Heschel as the proof that 'we cared about the civil rights movement.'" And even in Heschel's time, she rightly noted, not all American Jews actively supported civil rights.

Of American Jewish and Black American history during the civil rights movement, she said, "It is the same history . . . Jews who are white would benefit from *really* learning that history."[37]

Anthony Russell, a Black Jewish Yiddish performer, said in a phone interview, "The fact that we have this 'example' of Jewish activity within the civil rights movement [has been] inspirational to any number of generations that came after that. It has given them a platform from which to build ongoing solidarity with the Black community. It has allowed them to observe possibilities and paradigms . . . that involve working actively on behalf of other people you're in community with."

But the other side of this coin is that, for many white Jews, "It's looked on as an almost passive concern. This is something that we can look back on and be proud of and perhaps can inspire us to continue the same kind of work, but perhaps not.

"It's both of these things," he said. "It's simultaneously an invitation to rest on one's laurels and something that provokes people to action."[38]

That history, and the relationship between white Jews and Black Americans, was arguably to become still more complicated in the second half of the 1960s.

The martial law under which Israeli Arabs lived was finally lifted in 1966. But with the Six-Day War in 1967, millions of Palestinian Arabs were effectively brought under occupation in the lands Israel seized, including the West Bank.[39] This was not necessarily out of keeping with a Zionist vision, if Zionism just means a Jewish state in a particular part of the world to which they feel and assert a historic connection. The US ambassador to Israel under Trump, for example, tried to get the United States to adopt the Israeli term for the West Bank, "Judea and Samaria," an invocation of biblical terms.[40]

But it was decidedly not in keeping with a liberal or democratic vision. "If we keep holding the territories," Israeli finance minister Pinchas Sapir reportedly warned shortly after the Six-Day War ended, "in the end the territories will hold us."[41]

Many in the civil rights movement were not waiting for "the end" to arrive to denounce Israel. There were already some who did not believe that white people—including American Jews—should participate in the civil rights movement. "Get off the bandwagon," Stokely Carmichael of the Student Nonviolent Coordinating Committee (SNCC) told white Americans in 1965[42] (Carmichael would, later in the decade, head to Ghana and then Guinea, take the name Kwame Ture, and campaign for a Pan-African revolution).[43]

Two years later, in 1967, roughly two thousand activists came together at the Conference for New Politics. The conference's Black Caucus put forth a resolution that said that the Six-Day War was an "imperialist Zionist war" and that the white nation of Israel had taken Arab land.[44] This, again, presumes that all Jews are white, which all Jews—and particularly all Jews in Israel—are not.

This is, of course, not to say that racism in Israel against Jews of color does not exist, or that Israelis themselves never fall into the trap of equating Jewishness with whiteness. Many did then and do now. The tens of thousands of Moroccan Jews who went to Israel between the late 1940s and early 1950s faced discrimination on arrival.[45] In fact, in the 1970s, Israel had its own Black Panthers movement, which was modeled after the American Black Panthers but was made up of Jews whose families had come from North Africa and elsewhere in the Middle East, and who wanted to push back against government discrimination.[46] Ethiopian Jews in Israel have long faced prejudice and racism.[47] And it is not only a matter of history: one interviewee who made aliyah, whose mother is Asian, told me of the discrimination she faced both from other Jews in the United States and also from Israelis after arriving and becoming an Israeli citizen.[48] Still, the fact remains that both Jews and people who are not Jewish have a long history of erasing Jews of color when thinking and speaking about Jews in the United States and around the world.

Jewish delegates expressed dismay and left the Conference for New Politics in disgust.[49]

To assert that this was the moment at which American Jews and the civil rights movement split is untrue, or at least an oversimplification. There were schisms before the 1967 war, and it is not as though all Jews abandoned civil rights in 1967. A 1968 poll showed that Jews were nearly twice as likely as white Protestants to consider "racial inequality" the major issue of the day and that roughly one third of Jews thought civil rights reforms were "too slow."[50]

Further, as Marc Dollinger argued in *Black Power, Jewish Politics: Reinventing the Alliance in the 1960s*, American Jews were not as scared off by Black power as this traditional narrative might lead one to believe. Jewish leaders tried to downplay the significance of the Nation of Islam, a Black separatist movement, noting that they did not speak for all or even most Black Muslims. A 1964 article in *Congress Bi-Weekly* by Shad Polier, a leader of the American Jewish Congress,

called Black power a reasonable response to the slow pace of progress. Polier wrote that the real concern was not Black antisemitism
but the "response of the Jew," whom he did not want to see alien-
ated from the fight for racial justice.[51] The important thing, in other
words, was that American Jews were both understanding and engaged. There was a society to fight for.

Nevertheless, something had shifted not only between movements
but within them. Some on the Jewish Left felt it, too. There were, of
course, always people who opposed Zionism, but it was the position
of an increasing portion of the American Left more generally after
the 1967 war that Zionism was not a struggle for the underdog but
rather a fight by an aggressive and imperialist power. Some Jewish
leftists—as opposed to liberals, the vast majority of whom were still
proudly Zionist—could not help but notice this shift, particularly as
they protested against the US bombing of Vietnam.[52]

Some Jewish progressives responded by throwing themselves more
actively and consciously into civil rights. Rabbi Steven Schwarzschild
in 1968 argued against those who said Jews should abandon liberalism and integration for more explicitly Jewish causes but did so on
Jewish terms. "I also feel constrained to say, as a Jew and as nothing
but a Jew, 'If that's the way it's going to go, then I can have no part
in it.'"[53] Schwarzschild similarly grounded in Judaism a leftist argument against violence: "Two thousand solid years of *actual* Jewish
history—and to hell with all the theorizing—is quite *unqualifiedly* de
facto the most extraordinary exemplification of persistent practiced
pacifism in the history of the human race," he said in 1967, adding,
"with the possible exception of what the American Negro community has been doing in this country in the last ten years."[54]

These leftist Jews also worked to hold Jewish institutions accountable to their own stated values. In 1968, Jews for Urban Justice (JUJ),
which was composed mostly of young professionals, many of whom
had worked with groups like the SNCC and Students for a Demo-

cratic Society (SDS), held a press conference at the Union of American Hebrew Congregations. There, they issued their forty-page *Report on Social Action and the Jewish Community*, which concluded that the Jewish community in Washington, DC, was broadly disinterested with addressing inequity. Just two of the almost three dozen synagogues in the Washington area had planned any activities related to urban poverty. The report issued a number of suggestions, including that rabbis and community leaders address head-on "the issue of dishonest and discriminatory practices among members of their own congregations." Local Jewish leadership acted with characteristic self-reflection, which is to say they took umbrage. "I take strong exception and objection to this group of self-righteous people," said Brant Coopersmith, area director of the American Jewish Committee. The survey was "a lot of paper based on very insufficient data," said Rabbi Eugene J. Lipman of Temple Sinai. Liberal activist Bernard Mehlman, who had sponsored JUJ meetings at his synagogue, said that the survey reflected "a kind of Jewish self-hate." Still, he urged JUJ to return to Temple Micah so they could consider the words of the prophet Isaiah, namely, "Come let us reason together."[55]

Perhaps, though, it was JUJ, not Mehlman, whose reason won out in the end. Temple Micah is the synagogue I joined in 2020. It now also sponsors Micah House, a transitional residence for women in recovery from substance abuse; runs food drives for local pantries; and, in 2021, put on a series of talks on the "systematic disenfranchisement of Black communities." It isn't that mine is a radical synagogue—it's a welcoming space that prides itself on progress, but it isn't particularly revolutionary. It's just that what was once seen as radical is now, in part because of the pushing and prying of radicals, comfortably in the American Jewish liberal mainstream.

The response from the Jewish Left came through deed as well as word. In 1969, Arthur Waskow organized the Freedom Seder, which was sponsored by JUJ and weaved the story of Passover, Black Americans, and the war in Vietnam together, and brought Jewish and Black Americans together, too. The idea was to explicitly marry the tenets of Judaism with the American political Left.

I asked Waskow whether this was the moment that the Black civil rights movement and American Jews split.

"For some people, I think it is," he told me, his beard long and gray, his head adorned with a skullcap embroidered with menorahs. And it wasn't, in Waskow's telling, only because of the 1967 war. "The official community got what it needed, in a narrow sense, in American society. The original energy of the leadership council for civil rights, the writing of the civil rights bills as they're proud to say—all that came when it was still a question whether Jews were really white people or not. It became clearer and clearer that Jews had become white people." (Baldwin put it similarly: "Now, since the Jew is living here, like all the other white men living here, he wants the Negro to wait. And the Jew sometimes—often—does this in the name of his Jewishness, which is a terrible mistake. He has absolutely no relevance in this context as a Jew. His only relevance is that he is white and values his color and uses it.")

But if that was the case for some people, then "for other people," it was not the time to split from the civil rights movement.

Waskow said the Freedom Seder was both the climax of Jewish activism in the 1960s and the foundation for a new radical Jewish movement. There were some for whom the late 1960s reaffirmed the path that should be traveled. The contestation of Jews' place in the civil rights movement, for some, reaffirmed that this was where they wanted to be, and a fight of which they wanted to be a part, and to which they wanted to bring their whole Jewish selves. Waskow believed that they took the first steps, creating the route that many more travel today.

Waskow also charged that the split between radical Jews and mainstream, purportedly liberal Jews was evident before 1967. There were "almost no organized Jewish opponents of the Vietnam War, except for radical folks who helped make Freedom Seder real," Waskow said. "Most of the Jewish establishment kept their mouths shut at best, and at worst supported Lyndon Johnson and Nixon," Waskow said. American Jews were, broadly speaking, opposed to the Vietnam War. But Waskow's allegation is that that broad support was not transformed into meaningful action by most of America's Jewish institutions. Still, Waskow noted, there were some prominent Jews who can be pointed to as having stood against the war. "Heschel opposed it," he said, "because he was a serious Jew."[56]

What is a serious Jew? Waskow was using it to suggest that Heschel was learned and devoted, and that that learning and devotion showed him that the Vietnam War was incompatible with Judaism. Still, there are surely others who, because of their devotion to Israel and fear that opposition to Vietnam would turn the administration against it, felt that their support for the war was the serious Jewish thing to do.

In any case, Waskow is not incorrect in saying that the kind of communal, activist, progressive Jewish life in which he and JUJ participated established a precedent for this type of Jewish activity. Still, by 1970, this iteration of the full-throated Jewish Left was unraveling. The group came out with what it felt was a balanced statement on Israel, which called for self-determination of Palestinians and Israelis, but some of its members were nevertheless smeared as Arab terrorist sympathizers. Opposition came from within as well as outside: members were confused and hurt as to why the group's obvious link to the Bund, the Socialist group referenced in chapter one, was kept hidden from them. Was it out of shame of the past? Shame of Yiddish? What's more, some felt, as time went by, that the group's commitment to radicalism was waning and that there was no place, at least now, for radical American Jewishness.[57]

Other American Jews spent the late 1960s and 1970s more openly embracing what could be called "Jewish rights" but is perhaps more accurately defined as Jewish nationalism. Dollinger, in *Black Power, Jewish Politics*, argued that many American Jews were less put off by Black power than they were inspired by it, incorporating it into their own movements.

Rabbi Israel Dresner, in a 1969 gathering on Black-Jewish relations, asked if Jews couldn't benefit from the "sudden thrust of the Black people." Now, he said, he saw the possibility of Jews walking into work with a "kaputah, a beard"—in other words, expressing themselves in a more instantly recognizable "Jewish" way.

Leonard Fein, a Jewish American writer and activist, asked that same year whether Jews were not the "unintended beneficiaries" of Black power politics, which would give "more elbow room for Jewish assertiveness."[58] He was apparently proved right in very short order: The next year, education consultant Donald Feldstein reported that it was no longer embarrassing to be in a student group with the word "Jew" in the name. And as Black students pushed for Black studies, Jewish students, too, pushed for explicitly Jewish studies. Some American Jews applied this new outward embrace of a Jewish identity to the counterculture: neo-Hasidic spiritual leader Rabbi Shlomo Carlebach observed in 1967 that young Americans were "reaching for their own ethnic identities." He founded the House of Love and Prayer in San Francisco, which presented Judaism in the terms of peace and love of the day.[59] Many American Jews became invested in the plight of Soviet Jews, campaigning for their right to leave the country—and doing so without fear that they would, in the process, be seen as disloyal to their own country, the United States.[60]

Other American Jews became, literally, violently Jewish. Rabbi Meir Kahane founded the Jewish Defense Corps, later known as the Jewish Defense League, in 1968. The Jewish Defense Corps/League was deemed to be comprised of "misguided zealots" by twenty-six

Jewish organizations in 1970. Though Kahane said, "Blacks deserve nothing from us and that is what they will get," he also allowed a parallel between the Black Power movement and his own. Both "use unorthodox or outrageous ways" to meet their ends. "On this, we don't differ. We don't differ on their wanting to instill in their young people ethnic pride. Not at all."[61]

Black power even made the cause seen as a main driver in the rift between Black and Jewish Americans—Zionism—easier for American Jews to push. Black nationalism and Zionism were different in substance, but the very existence of the former lessened concerns around "dual loyalty," an old trope that suggested Jews were never really loyal to the country of which they claimed to be a part.

In other words, these American Jews hadn't been pushed away by identity politics. Rather, they had embraced identity politics for themselves. Or, as a 1973 article in *Tradition* by Professor Ronald Rubin articulated, young Jews hadn't abandoned liberalism; they had replaced it with Judaism.[62] (Ironically, this was also around the time that Evangelical Christians came into their power as a political group with the rise of televangelism. They were courted by President Jimmy Carter, who was himself a born-again Christian, though his liberal policies eventually turned off Evangelical voters.[63] *Newsweek* put Evangelicals on their cover in 1977.[64] Evangelicals had their own relationship to Israel, which many of them believe is necessary to fulfill the end-times prophecy.[65] This is to say that, at the point in American history at which Jews became comfortable claiming explicitly Jewish political issues, at least one of those issues—Israel—was transforming from a "Jewish issue" into something else in the American political landscape. Incidentally, this corresponded with the 1977 election of Menachem Begin and the first right-wing government in Israel.)

Some Jews, then, did explicitly reject liberalism and universalism at this point in the history of the civil rights movement. Susannah

Heschel took a more measured view about what was happening at this time. Both Israeli victory in the 1967 war and Black nationalism gave a boost to Jewish nationalism, she told me. To her mind, both groups had considerable accomplishments. Israel took "millions of refugees who arrived with nothing, some of them straight out of death camps, and [turned] it into a state," and the Black Panthers established education and welfare programs. "I don't think that should be a cause for some kind of concern that we don't care about each other."

Black and (white) American Jews, she said, "moved on parallel tracks, doing our own nationalist thing. My father, I think, recognized that and accepted it. That that's what's happening right now."[66]

Even the American Jew most visibly committed to civil rights realized, in the words of his daughter, that times were changing and that the civil rights movement and relationship between Black Americans and white Jewish Americans wouldn't be what it was before.

This bifurcation in the 1970s and '80s was less comfortable for biracial Jews. In *The Soul of Judaism: Jews of African Descent in America*, sociology professor Bruce D. Haynes quotes a Black Jewish man called Seymour, who had been involved in both a Zionist Socialist youth movement and graffiti subculture in Harlem. Seymour said, of his experience growing up in the 1970s and '80s, "I would sort of defend my Black brothers from any sort of attack that might be based purely on the color of their skin. . . . And to Jews in the face of Blacks who would be antisemitic, you know, my perspective was that, 'Do you like me? I'm Jewish, so what do you really know? How deep is this feeling and what's up with that?'"[67]

Another part of this equation is that it is not only Black power that made (white) American Jews comfortable asserting their Jewishness explicitly, specifically, and politically, or embrace Jewish identity politics. It was also whiteness.

The assimilation of many Jews into white, upper-middle-class

America, per Baldwin and Waskow, complicated the relationship between American Jews and the civil rights movement. It also created the conditions that allowed American Jews to feel comfortable asserting their Jewish identity more fully. Jews who were worried about being able and allowed to create a life for themselves and their children in America and didn't want to seem ethnic or different or be seen as only being concerned with other Jews; Jews confident of their place in America could and did.

This was true culturally as well as politically. There is perhaps no clearer example than *Fiddler on the Roof*, which celebrated and romanticized and created an outlet for a kind of nostalgia for the shtetl. This was a far cry from the carefully sanitized renditions of Jewish families on television in decades prior. The 1964 musical was a smash hit and was turned into a movie in 1971.[68]

In the interim, in 1969, a group of Black and Puerto Rican middle school students put on the show in Brownsville, Brooklyn. Black-Jewish tensions were high in and around the school, with the Jewish teachers on strike. (In 1968, a radio talk show host by the name of Julius Lester decided to go on air and read a poem by a fourteen-year-old student at Ocean-Hill Brownsville that went, "Hey, Jewboy, with that yarmulke on your head, You pale-faced Jewboy, I wish you were dead." In a twist, Lester went on to convert to Judaism, and, in 1988, won the National Jewish Book Award for his memoir, *Lovesong: Becoming a Jew*.)[69]

In this context, the drama teacher, Richard Piro, hoped the show would help. Some in the school and community thought the students were going to make fun of Jews and tried to halt production, going so far as to let the producers know that the school was putting it on without permission. In response, permission was granted, and members of the creative team—Jerry Bock, Sheldon Harnick, and Joseph Stein, as well as producer Hal Prince—traveled to Brownsville to see the show.[70]

The students did not make fun of Jews in their production. They related to them. Their parents also wouldn't let them date the boys they wanted to; their families were also facing eviction, just like Tevye and his clan did in fictional Anatevka.[71]

And here, too, one can ask: Who, in this story, are the Good Jews or Bad Jews or Real Jews? Those who claimed ownership of *Fiddler on the Roof*? Or those who gave permission to the students, letting them act out the story? Alternatively, one could extract from this anecdote that there are different ways to view protecting Jewish stories. One is to insist that they remain for and by Jews. The other is to share them and hope that others can relate to them and see themselves in them, too. The tale of Anatevka in Brownsville is one of the universalist versus particularist debate between American Jews, an argument over whether being a Good Jew means concerning yourself with all people or whether it means focusing on your people, the Jews.

The focus on issues of explicitly Jewish concern, made possible in part by Jews' more comfortable position in American society, was also reflected in the conversation around intermarriage, and, relatedly, women's rights. Intermarriage was once seen as a litmus test for whether Jews were capable of assimilating. Jews had assimilated. And now Jews were intermarrying, too.

Interfaith marriage rates—likely due to a combination of more liberal social attitudes, an increased number of women in higher education, and counterculture as a space where people connected with like-minded individuals of various religions—went up in the 1960s and '70s. The proportion of Jews who intermarried nearly tripled. Before the 1960s, it was just 6 percent; from 1961 to 1965, up to 17.4 percent; from 1966 to 1972, it was up to 31.7 percent. Individual Jews, then, were more open to intermarrying, and qualitative research conducted suggests not only that Jewish families were more open to intermarriage but that intermarried Jews began to more consciously consider how to raise their children as Jewish and how to create Jewish families.

This did not assuage the concerns of mainstream Jewish organizations and institutions. As one activist put it in the 1970s, intermarriage and what to do about it became the "single most pressing" issue for the organized Jewish community[72]—and for the sociologists and demographers they hired.

Sociologist Erich Rosenthal fretted over the low fertility rates of American Jews and increasing rates of intermarriage.[73] "Does this mean extinction?" John Slawson, executive vice president at the American Jewish Committee, asked on reading Rosenthal's data.[74]

There were other studies, and other pieces of information, that could tell a different story. "To have withstood Hitlerism and still create Israel . . . it is utterly premature to be morbid about intermarriage," a rabbi from the Society for the Advancement of Judaism in New York offered in 1964. And in 1966, sociologists Sidney Goldstein and Calvin Goldscheider conducted a study of Jews in Greater Providence, Rhode Island, and found that, while the rate of intermarriage was higher, so, too, was the rate of conversion of the non-Jewish spouse. What's more, they found that most children from intermarriages were, in fact, raised Jewish.[75]

This was not enough to challenge the prevailing narrative. In 1970, that narrative was amplified dramatically. By the 1970s, there was a powerful, well-funded effort to study Jewish families under the term "Jewish survival."

In 1970, the Council of Jewish Federations and Welfare Funds sponsored the first National Jewish Population Survey. The NJPS found that 31.7 percent of Jews who married between 1966 and 1972 chose a non-Jewish spouse. In 1976, the American Jewish Committee sponsored a national study. Sociologists later summarized the findings from that study: "Most non-Jewish spouses do not convert to Judaism; the level of Jewish consent and practice in mixed marriages is low; only about one-third of the Jewish partners in such marriages view their children as Jewish; and most such children are exposed

to little by way of Jewish culture or religion." Most rabbis were dis-
couraged from performing intermarriages—the Central Conference
of American Rabbis officially declared "its opposition to participation
by its members in any ceremony which solemnizes mixed marriage"
in 1973—and rabbis who argued that they were alienating members
of their own community and potential converts by taking this stance
were evidently in the minority.[76]

Thanks to a combination of belief that it was primarily a woman's
responsibility to raise a family and the halakhic rule of matrilineal
descent that says that Judaism is passed on through the mother, the
responsibility for saving the family, and blame for not doing so, fell on
Jewish women. It was Jewish men who dictated the contours of this
particular part of Jewish rights, and they had no qualms at doing so at
the cost of women's empowerment, or even women's dignity. Milton
Himmelfarb of the American Jewish Committee complained in 1975
that Jewish women were "contraceptive virtuoso."[77] In 1980, Rabbi
Harold Kushner argued that women in the workforce were part of
the deterioration of the Jewish family.[78]

One sociologist involved in producing these studies for Jewish
organizations was Steven M. Cohen.

"These Jewish organizations, run by men, give him [Cohen] millions of dollars to do these demographic surveys, where he comes
up with conclusions that confirm all of their biases, that women are
not having enough babies," Susannah Heschel, who is the chair of
the Jewish Studies program at Dartmouth, told me when I asked her
about how differently feminism and the civil rights movement were
perceived by American Jews. "In the 1970s and '80s, instead of creating and giving subsidies to daycare centers, making it possible for
women to have kids as well as a career, women were told, 'Don't have
a career. Wait until your kids are all grown up.'

"For a big generation of women, Steven Cohen really hurt us. Profoundly."[79]

Cohen was a leading figure in shaping how Jewish women and Jewish continuity were discussed. In 2018, he was accused of having spent years harassing female colleagues and subordinates. He did not deny the allegations. In 2021, it was revealed that, despite this, he'd been having private, off-the-record conversations with Jewish institutions and leaders about Jewish life.[80] There were still some in positions of power within Jewish institutions who felt that this man, who had effectively told Jewish women how to live professionally while harassing them privately, was worth listening to about what it meant to live a Jewish life.

In fairness, though, it was not only Steven Cohen. It was Jewish organizations, Susannah Heschel said, that cared about reproduction but not the women doing the reproducing. "So I don't count but I should still produce babies," she said, still, all these years later, obviously disgusted. "For what? For a world that doesn't care about women? Why would I want to do that?"

The studies weren't only used to dictate to Jewish women how to be. They also became devices through which Jewishness was measured. Beginning in the 1980s, the studies started using Jewish ritual as proof of one's Jewish devotion. How often do you light the candles, for example.

"They didn't ask about the subjective dimension. 'What does it mean to you? Why do you do this? How does it feel to you?'" Susannah Heschel said. "Nobody really wanted to know. Because, decade after decade, they didn't ask.

"What you can learn from these studies was quite limited," she said. "But they satisfied the people who paid for them."[81]

There was, then, a counterculture movement, and a radical Jewish movement, and a mainstream Jewish movement that nevertheless used the language of radical rights to agitate for issues they thought were directly of interest to themselves as American Jews.

But there were other American Jews, too, for whom the post-1967 change in the civil rights movement also cemented their own shift on the American political spectrum. Theirs was a minority view among American Jews, but it was one that would become one of the most influential. These individuals would go on to claim that theirs was, or at least should be, the majority opinion.

They were the neoconservatives.

Chapter 5

Right-Wing Jews

C ERTAIN PROMINENT INDIVIDUALS IN the so-called American Jewish community characterized 1967 as a breaking point for them. They had been liberals, these individuals said, but now their liberalism was being tested by civil rights and by the lack of space within the civil rights movement for Jews.

These people were part of a group that was known, at least for a time, as neoconservatives: free-market capitalists and foreign hawks who had grown disenchanted with the liberal project, or at least with the Democratic Party.

Not all neoconservatives were Jewish, but many were. From the beginning, there were conservative, traditional, and at points downright discriminatory strains of thinking within it, which evolved over time to be a minority within the American Jewish minority, but an exceptionally powerful one.

In 1952, Irving Kristol wrote an article for *Commentary* attacking liberals, who, unlike Joseph McCarthy, were insufficiently critical of

Communists for Kristol's liking.[1] He had written it at the urging of his editor, Elliot Cohen, who wanted his writer to demonstrate that Jews could be relied upon as Cold Warriors.

Kristol, along with Norman Podhoretz, is one of the two fathers of the formal movement that came to be known as neoconservatism.

Neoconservatives actually started out as leftist radicals. They were disciples of Leon Trotsky and, in the run-up to World War II, concerned themselves not with the fate of Europe's Jews, but with inter-Communist squabbles. On campus at City College in New York, the Trotskyists of Alcove 1 would battle it out with the Stalinists of Alcove 2.

Max Shachtman, a Jewish immigrant from Poland, is generally presented as the forefather of the neoconservatives. Eventually, Shachtman would be that rare Jewish Vietnam War supporter, but in the interwar period he was a Trotskyite. He actually met his hero but, in 1937, was expelled from the Socialist Party. Undeterred, he founded the Young People's Socialist League, which went on to become a training ground for people like Kristol, who was inducted into Trotskyism by the Socialist critic and editor Irving Howe. It was a move that Howe, who remained a Socialist, would come to regret.[2]

The neoconservatives were shaped by cultural influences and giants, too, like the *Partisan Review* and Lionel Trilling,[3] the literary critic and mentor to both Cohen and Podhoretz, who took over *Commentary* after Cohen—and who, along with Kristol, known as the movement's "godfather,"[4] was a leading neoconservative.

Why did these young people give up their affinity for Trotsky and Socialism? What happened to them? The same thing that happened to American Jews more generally: World War II happened. Stalin was well and truly in charge of the Soviet Union, and a Jewish state was recognized. The Trotskyites had been anti-Stalinist; now, in the postwar period, with Stalin solidly in charge of the Soviet Union, they were anti-Communist. And if the country best positioned to counter the Soviet Union appeared to be the United States, well, they would align themselves with the power of the American state.

But they remained radicals. They were not content with liberal values and containment of the Soviet Union. If they were to be against Communism, then they would do battle against it. They were Cold Warriors.

Eventually, neoconservatives would be associated with the Republican Party. But in the 1950s and '60s and even into the 1970s, they were still trying to save the Democratic Party from itself.

Podhoretz would say that the trial of Adolf Eichmann in 1961 was, for him, the ideological turning point.[5] The war in 1967 and increasing liberal criticism of Israel was significant, too. As was the civil rights movement: that *Commentary* is the place that published "My Negro Problem—And Ours," in which Podhoretz uses his personal childhood experience to express ambivalence with the civil rights movement,[6] suggests that at least some in this cohort were not as supportive of civil rights as the majority of American Jews.

Still, one wonders if this was as much an excuse as it was a reason. In 1963, Glazer (who was considered a neoconservative, though he did not accept the label) and Moynihan came out with *Beyond the Melting Pot*, which both essentialized Jews and juxtaposed them with Black Americans.[7] It was the same year that "My Negro Problem—And Ours" came out (the conclusion of which is that only miscegenation could bring about full integration, which was denounced by Jewish leadership not because the idea that marriage can end racism was wrong, or because it abandons productive policy and turns progress over to reproduction, but because they opposed its call to intermarry).

This was before the 1967 war. As tensions mounted in the civil rights movement, some of the neoconservatives were early to the argument that Jews should abandon civil rights in favor of what they deemed to be explicitly Jewish causes.

Perhaps this was because this generation, and particularly those involved in this specific intellectual tradition, forged in disappointment and disillusionment with Communism, the hardships of immigration and assimilation, and the tragedy of the Holocaust, genuinely feared for their own survival.

And perhaps it was because they believed the story America told itself about its own meritocracy, and that its white elite sat at the top because they belonged there. And believed that they should be there, too.

A combination of both reasons is, of course, possible, and even likely. It is both too easy and also incorrect to say that these Jews were forced out of the civil rights movement. Years before 1967, they were making points that stood in direct opposition to the civil rights movement. They were forging another, different path for themselves and were focused on a new path for American Jews.

Some—notably, Jacob Heilbrunn, author of *They Knew They Were Right: The Rise of the Neocons*—have argued that neoconservatism cannot be separated from the Jewishness of its fathers and followers.[8] Certainly, for some, the dual experiences of World War II and the foundation of Israel were personally significant and transformational.

Heilbrunn's argument wasn't universally accepted, and certainly not by those whom it was describing.

Douglas Feith, a neoconservative who served in both Reagan's and George W. Bush's administrations, does not credit the idea that neoconservatism was inherently or especially Jewish. It is wrong, he says, to think of neoconservative thought as "an intellectual strain of Judaism." Ideas and debate are central to Jewish culture. "Jews aren't big on catechism," he told me in 2019, "rather they are big on argument." He warned against ascribing a particular political school of thought to Judaism, arguing that Jews cite various Jewish principles to justify a wide range of political views. Some cite traditional Jewish ideas in support of neoconservative thinking, and some in opposition to such thinking. The same is true of Roman Catholics, he said, some of whom were prominent neoconservatives and others prominent anti-neocons.[9]

But whether neoconservatism itself is Jewish, some—many, even—neoconservatives were. In the postwar period, some Jews were good liberals, and some Jews were agonizing over their newfound wealth and status, and some Jews were scared of being called Communists, and some Jews were civil rights supporters, and some Jews were also, at the same time, criticizing liberals and arguing that they belonged in the middle class and calling for aggression against Communism and shrugging over civil rights. And all were resting on the Jewish tradition, or, rather, if they wanted to, all could point to some Jewish person or event involving Jews that had happened at some point in the past and call it tradition, or precedent, and claim that, see, this was the right way to go forward. The Jewish way.

In 1972, Podhoretz published another *Commentary* essay. This time, he argued that American Jews should concern themselves more with the age-old question, "Is it good for the Jews?" That he wrote such an essay spoke to how far Jews had come in the United States: once consumed with worry about assimilation and acculturation, and then largely (at least in theory) occupied with the project of advancing rights for all, American Jewish public intellectuals now felt enough security in their own place in society to demand a conversation that drew attention to what they perceived as their own communal distinctiveness.[10]

The problem with the question, "Is it good for the Jews?," though, is that it presumes that there is one answer. Indeed, it presumes that all Jews feel the same about what is good and bad for them, or at least that there is enough uniformity among Jews that something can be decidedly good or decidedly bad for them.

But by asking the question, Podhoretz and *Commentary* were claiming for themselves the right to answer it.

Commentary was launched as the journal of the American Jewish Committee in 1945. It was launched by Elliot Cohen, who had made a name for himself at *Menorah Journal*,[11] and who sought to make *Commentary* a great American Jewish publication. He brought on Nathan Glazer and, soon after, Irving Kristol.[12]

The American Jewish Committee made clear through a disclaimer that, though *Commentary* was an AJC publication (and indeed would be until 2007), the views in the magazine did not necessarily represent the views of the American Jewish Committee. Still, that it was the magazine of arguably the most elite, most establishment Jewish organization—one that had consciously claimed to represent what it deemed to be the best interests of the Jewish community—allowed those writing for it to write from a certain position of Jewish power and authority. It was a position of which *Commentary* authors took full advantage.

Commentary did not only publish strictly political musings. It also included assessments of theological thought[13] and literary stories.[14] But it was arguably above all the mouthpiece of an increasingly powerful group of Jewish thinkers.

As author and former *Commentary* associate editor Benjamin Balint outlines in his book *Running Commentary*, in the 1950s, *Commentary* writers, many of whom would come to be thought of as neoconservatives, styled themselves as liberal anti-Communists. That meant that this Jewish journal, associated with a preeminent Jewish organization, declared that to be an American Jew was to be anti-Communist (here, again, one can refer to the Lucy Dawidowicz essay on the Rosenbergs, in which she argued that executing Julius and Ethel Rosenberg was actually good for the Jews). *Commentary* kept careful track of the Soviet Union's crimes against Jews (like the

Doctors' Plot, in which Jewish doctors were accused of plotting to poison Soviet leaders, and the execution of Yiddish writers). Cohen maintained the magazine's hard line after Stalin's death in 1953. They rang out their warnings for the threat of Communism at home as well as abroad.[15]

Some former friends of the magazine charged that it had ulterior motives in doing so.

"My complaint against *Commentary*," Irving Howe wrote in 1954, "was not that it ceased to be socialist; it never had been that; but rather that it has become an apologist for middle-class values, middle-class culture, and the status quo." (Howe went on to found his own Socialist magazine, *Dissent*.)[16]

But Kristol and Glazer were not only defending middle-class values; they were defining them for American Jews.

Elliot Cohen, overcome by anxiety, took his own life in 1959 at age sixty.[17] He was succeeded as editor of *Commentary* by Norman Podhoretz.

Six years into the role, Podhoretz said, "*Commentary* today is less concerned than it used to be with the sociology of the American Jewish community, with their social sciences in general, with the character of the American middle class. It does, however, remain as interested as ever in the meaning of the US in international affairs, the importance of ethnicity and religion in American life, the quality of contemporary culture, the problems of education, the elusive nature of Jewishness."[18]

This intellectual set resented the new, young radical youth, whom he called "Know-Nothing Bohemians" and "spiritually underprivileged." *Commentary* felt that the so-called New Left was anti-American and took it upon itself to defend America. Podhoretz believed it his duty to convince Jews "that radicalism was their enemy, and not their friend."[19] Radicalism, in other words, was bad for the Jews.

Podhoretz brought onto the team Milton Himmelfarb, Irving Kristol's brother-in-law, who, in 1969, wrote that American Jews earned like Episcopalians and voted like Puerto Ricans. "Jewish students will occupy a university building," he said of campus unrest in the 1960s, "or approve its occupation, to support the demands of Blacks, but not to prevent or abolish quotas that hurt Jews."[20] There were Jews among the "radicals," criticizing the United States. And there were, clearly, Jewish voters and Jewish students who believed that pluralism and universal rights were in their own interest. But Podhoretz and Himmelfarb did not see it this way.

Podhoretz also came to argue that "hostility to Israel often spills over into a hostility to Jews, just as a hatred of middle-class values spills over into the hatred of Jews," and that this was "despite the fact but also of course because of the fact that so many of its members are Jews."[21] Jews of this political persuasion, this argument holds, give cover to antisemitism on the left. Podhoretz determined that antisemitism on the left had become more dangerous than that on the right.

This combination—of radicalism they deemed unsavory, of hostility to Israel that they believed crossed all too easily into antisemitism, and of anti-Americanism by which they could not abide—convinced the *Commentary* crew that the liberal coalition was no longer a viable home for them. Soon, they would make the same case about the Democratic Party.

In the 1972 presidential election, several members of the *Commentary* staff, including Podhoretz, voted for Richard Nixon, so convinced were they that McGovern and his supporters were, per Podhoretz, "hostile to the feelings and beliefs of the majority of the American people." Sixty-five percent of American Jews, on the other hand, voted for McGovern (this did not help McGovern, who won only 17 electoral votes).

The *Commentary* crew supported Democrats in the mold of Daniel Patrick Moynihan, the future senator from New York who co-authored *Beyond the Melting Pot* with Glazer, and Henry "Scoop" Jackson. After 1976, in which Jimmy Carter defeated Moynihan in the presidential primaries, the *Commentary* writers and editors more openly embraced a new conservatism.[22]

This set of thinkers decided to make what they felt was a more intellectually meaningful conservatism. In doing so, some saw it as taking on a new and exclusively hawkish approach to foreign policy. Literary critic George Steiner noted that the magazine stopped publishing him, which he assumed "arose from both my doubts about Vietnam and my deepening fear about the development of Israel's policies and society." American author Edward Hoagland felt *Commentary* was "banging the drum for the Cold War and fighting the Arabs."[23]

Podhoretz brought on new writers, like the scholar Ruth Wisse and then Harvard law student Elliott Abrams, who later served under Presidents Reagan, George W. Bush, and Trump (and some who were not Jewish, like political scientist Jeane Kirkpatrick).[24] By 1980, the transformation—insofar as it was a transformation and not a continuation—was complete. "For the blunt and crippling truth," wrote executive editor Neal Kozodoy that year, "is that the Jews of this country, especially those Jews who speak for the interests of their fellow Jews and who are appointed keepers of Jewish conscience, have not yet fully developed the habits and attitudes appropriate to free men."[25] Jews, Kozodoy wrote, had been too obsessed with the Jewish commitment to help others.

Leaving aside that many Jews were more interested in saying they wanted to help others than they were in helping others, the reality was that, in 1980, most Jews were still voting for Democrats. Most Jews had not abandoned liberalism. Were most Jews, then, according to Kozodoy, not free?

• • •

In 1980, Ronald Reagan was elected president. Here was a president whose worldview matched that of the neoconservatives: social conservatism, free market economics, and, crucially, Cold War
hawkishness and a belief in the inherent and special goodness of
America. In Reagan's United States, as in the minds of the neoconservatives, might should be married with moral right. More American Jews voted for Reagan than for any other Republican president
since Abraham Lincoln: 39 percent.[26] It's a significant number. But
39 percent is still not a majority. In the run-up to the 1984 presidential
election, the Republican Jewish Coalition called itself the National
Jewish Coalition, and claimed it was a bipartisan organization "that
will demonstrate to fellow Jews that genuine and immediate Jewish
interests are at stake . . . and that these interests are best served—and
best protected—by the re-election of the Reagan team."[27]

Still, in 1984, 68 percent of American Jews voted for Walter Mon-
dale over Reagan.[28] Lucy Dawidowicz wrote in *Commentary*, "The
lopsided Jewish voting pattern resembled that of the blacks, the un-
employed, and persons in households earning under $10,000 a year,
even though Jews in no way resemble those groups or share their
social and political interests."[29]A decade and a half after Himmelfarb
said Jews vote like Puerto Ricans, Dawidowicz, too, seemed perplexed that two thirds of American Jews would vote for the party
that was openly committed to ensuring a pluralistic United States.

The magazine cheered Reagan on throughout his time in office, and
in particular early on in the Reagan years, before Reagan decided to
try to work with Soviet leader Mikhail Gorbachev.[30]

Feith said that, at this time, neoconservatism effectively became
Reaganism, and that it no longer existed (if it ever really did) as a distinct school of thought about foreign or domestic policy.[31]

Some of the *Commentary* crew—like Elliott Abrams and Carl Gershman and Robert Kagan and the decidedly not-Jewish Moynihan and Kirkpatrick—even went to work for Reagan. And both those who wrote of Reagan (and were read by members of the administration) and those who worked for him made the case for American military muscle. So committed to this particular project was *Commentary* that historian Ronald Steel said in 1975, "Relax, *Commentary*. Just because the US has not bombed or invaded anyone this week does not mean that Western civilization is teetering on the brink of ruin."[32]

In 1984, *Commentary* argued for "Why an Arms Build-Up Is Morally Necessary."[33] Throughout the decade, the magazine argued against arms control, with Podhoretz deeming "arms negotiations with the Soviet Union" to be "a lie and a fraud and a deceit to the nation."

There were some disagreements among the neoconservatives on matters of foreign policy—Kristol believed that human rights was a left-wing project; Podhoretz thought threats needed to be articulated so that the public would support military programs; and Abrams, who was assistant to Secretary of State George Shultz, was on what Heilbrunn called a "democratic crusade"—but as a whole they were essentially on board with the basic project of peace through strength.[34]

And so began a new, though not the last, chapter of "Jews cause all the wars."

The trope of Jews being responsible for all of America's wars did not begin with the neoconservatives and Reagan. The trope that Jews had orchestrated World War I provided antisemitic undertones to the America First and isolationist project pushed by Charles Lindbergh and his ilk. This then morphed into the fear that US entry into World War II would be seen as once again going to war for the Jews (notably, some prominent American Jews held and voiced this concern). And yet another iteration gained favor later, after the disastrous war in Iraq.

Still, one must acknowledge that the contemporary version of the trope that Jews cause America's wars begins with Reagan and the neoconservatives. For one thing, many of the neoconservatives did make it into positions of influence in the administration. And where previous prominent American Jews in administrations hadn't been able to push policy that helped Jews (like Morgenthau) or were profoundly and openly disinterested in doing so and spoke of realpolitik (like Henry Kissinger, Richard Nixon's secretary of state), the neoconservatives tried to make moral cases for the policy they were pushing, including, in some cases, that their way—particularly in supporting Israel and fighting Communism—was good for the Jews.

The trope only became more pronounced during and following George W. Bush's administration, as America came to review its recent history.

The neoconservatives were generally out of favor in the early 1990s. William Kristol, Irving Kristol's son, found his way into the elder Bush's administration,[35] but that presidency was defined by realpolitik. And though they generally agreed with President Bill Clinton about intervening during the war in Bosnia, one could argue that at least some neoconservatives were still stuck in the 1980s: in 1993, Irving Kristol wrote an essay in *National Interest* under the headline, "My Cold War."

"There is no 'after the Cold War' for me," Irving Kristol wrote. "So far from having ended, my cold war has increased in intensity, as sector after sector of American life has been ruthlessly corrupted by the liberal ethos."[36]

But the neoconservatives, if indeed that title still applied (some believed it was outdated; Podhoretz in 1996 declared the movement dead, arguing that it had successfully become "conservativism"),[37] did not stay on the outskirts or in the past for long. Some reemerged in the halls of power during the younger Bush's time in office. Of Bush, Podhoretz wrote, "As a 'founding father' of neoconservatism who had broken ranks with the Left precisely because I was repelled by its 'negative faith in America the ugly,' I naturally welcomed this new patriotic mood with open arms."[38] Charles Krauthammer wrote in *Commentary* that the Bush administration was "neoconservatism in power."[39]

Unfortunately, that power, following the terrorist attacks of September 11, 2001, called for spreading democracy and fighting terror, which in turn led the United States into the ill-fated Iraq War.

Some, today, consider "neoconservative" to itself be an antisemitic term, so closely associated is it with this idea that Jews started all the wars. In 2017, former CIA operative Valerie Plame Wilson tweeted, "American Jews are driving America's wars." She linked to an article with those words as the headline. Plame Wilson, as if offering a defensive explanation for trafficking in this trope, then added that she was "of Jewish descent" but "many neocon hawks ARE Jewish." (She then apologized, saying she'd overlooked the problematic parts of the piece because she "zeroed in on the neocon criticism.")[40]

The idea that neocons cause wars and Jews are neocons and therefore Jews are the reason the United States gets entangled in wars is, for some, a tantalizing one. It promises to help you forget that Reagan wasn't Jewish, and George W. Bush wasn't Jewish, and that Israel was supported not only by Jews (and is indeed backed by the Evangelical right), and that the idea that the United States pursued a foreign policy out of selfless altruism because some members of a religious minority said so ignores the fact that US foreign policy, even when dressed up with morals, is always self-interested. It offers to let people believe that foreign policy misadventures and failures are not about imperialism, or America. They can be instead about the neoconservatives. About the Jews.

There were some neoconservatives among the policy makers who pushed for the war in Iraq, and some of them were Jewish, and they may even have been neoconservatives because of their American Jewish background, shaped as it was by history and concern for Israel and desire to belong to a certain segment of American society. But Jews are not responsible for America's wars. America is.

Back in the 1980s, American Jews were themselves aware that this could neatly be seized upon by antisemites, bringing to the fore that old trope of dual loyalty, and how easy it would be for people to assume that American Jews were first and foremost concerned about Israel, and not the United States. For example, Douglas Glant, one of the founding members of the Republican Jewish Coalition, wrote to Max Fisher, a major philanthropist and political donor, "you should know that most of the young Jews I will be bringing to the Party are very pro-Israel but most of them consider themselves Americans first, and are concerned with such things as domestic national security, economic stability, fair taxes, crime, schools, etc., in addition to Israel."[41]

Reagan, of course, was a born-again Evangelical, and Christianity was central to his political project. But this did not upset the neoconservatives; if anything, it was useful to them. "The fact that the Moral Majority is pro-Israel for theological reasons that flow from Christian belief is hardly a reason for Jews to distance themselves from it," Irving Kristol wrote in 1984. The real threat to Jews and Israel were liberal, secular Jews. Of the Christian right, he said, "It may be their theology, but it is our Israel." The idea that they were being used by, and not using, deeply Christian conservatives evidently either did not occur to Kristol or did not bother him.

In the early 1990s, both Podhoretz and Midge Decter would defend Pat Robertson and the religious right from charges of antisemitism; after all, he supported Israel. When the ADL denounced Robertson as an antisemite, the former argued that his support for Israel mattered more than any antisemitic undertones of comments he may have made; the latter assembled a letter signed by like-minded Jews asserting that it was the ADL that was in the wrong, for it was acting as though Judaism and liberalism were the same thing.[42]

There were many lasting effects of the neoconservatives. The Bush era, in which might makes right was morphed before the country's eyes into might makes might—moral arguments becoming more and more stretched, less and less legible, but offered up all the same—first in the run-up to and then during the war in Iraq, was one. This was another. Right-wing Jews could excuse and dismiss and provide cover for antisemitism at home in exchange for support of Israel. They could say that someone wasn't an antisemite, and that, if they were, it didn't matter, because they supported Israel. This was a long-lasting inheritance left by this cohort of right-wing Jews for their successors. It was one that would manage to make it out of the twentieth century and enter into the twenty-first.

Podhoretz suggested that Americans associated Jews with the middle class, and so sometimes denigrated the middle class as a way to denigrate Jews. So, too, did he argue that it was, in a sense, the responsibility of Jews to defend the American middle class and the values of the American middle class from radicalism, which he identified as the main threat to Jews in America (notwithstanding the fact that many of these radicals were themselves Jewish).

The neoconservatives argued that Jews should take and defend their place of belonging not only in America, or in white America, but in a very particular section thereof: white, middle-class America. And that they, the neoconservatives, should speak for the Jews of this segment of America in which they most decidedly belonged.

But this is separate from (though related to) the question of how American Jews came to be not just a part of but a synonym for the American middle class—and not only the middle class, but American finance, American wealth, American money, and a related question: How can the cliché of Jews and money be separated from the real story of American Jews, which is one of both fighting for and profiting off of labor?

Chapter 6

Laboring Jews

THE FIRST PERFORMANCE OF William Shakespeare's *Merchant of Venice* took place in 1605, which means that it was over four hundred years ago that audiences first met Shylock, the Jewish moneylender who is out for a pound of flesh after the titular merchant defaults on a large loan.[1] Shrewd audience members would have noted that Shylock says that his cruel behavior toward others reflects how society treats him, a Jew, as less than human; the rest of the crowd would have just seen the moneygrubbing Jewish stereotype with which they'd have already been long familiar.

The stereotype of Jews not just being obsessed with money, but controlling and manipulating the financial systems, goes back centuries. In the United States, even in the years after Jewish mass immigration to America, Henry Ford published an updated version of *The Protocols of the Elders of Zion* and believed and pushed the idea that Jewish people and their money had too much influence.[2] American Jews were keenly aware of the stereotype, pointing to romanticized versions of the shtetl and the Lower East Side as more "authentic" times in Jewish history.

American Jews were joined in their awareness of this stereotype by the rest of America.

"Get back in your Cadillac," my father recalls being told on senior skip day in high school, "and go back to Israel."[3] Forty years later, when I was in middle school, a boy put a dime down in front of me. When I didn't pick it up, he said, "You passed the Jewish test." I can still remember looking down at the coin, lying there on the round black table, wishing it would disappear into thin air and maybe take me with it.

In high school math class, another, different boy, looking at aphorisms on a bulletin board, read, "A penny saved is a penny earned," and then said, "Who said that? I bet they were Jewish" (the phrase is actually believed to have originated with Benjamin Franklin, who was Quaker, but when I went to explain that, and why that kind of joke had historically been weaponized against Jews, the teacher brought our attention back to some calculation I no longer remember).

"I got coins thrown at me," my friend Ed Delman, another of the few openly Jewish kids in my grade, told me. That was the most aggressive example, he said, of a mentality that was "so ingrained in the community."[4] My younger brother confessed that he, too, got pennies thrown at him by his swimming teammates. My younger sister told me that what upset her most as a child in that town—that town where she knew that Jews were different and all she wanted was to be like everyone else—was the stereotype that Jews were cheap. She wanted to dress as though she were "anti-cheap," she said. (I noted to her that there was also the stereotype of the materialistic Jewish woman. She agreed but said that our peers weren't as aware of that one.)[5]

And it wasn't just my town. I am sure that many children who went to school with lots of people who were not Jewish never knew the torment of the coins, but, for others of us, it was part of growing up Jewish amidst people who weren't. Joshua Fitt, who now works as an analyst in Washington, DC, but who grew up in Buffalo, told me, "Back in high school, I'd randomly notice a coin had been thrown at me some way or another. I'd always pick it up. I think in some ways that fed into what the person was trying to do, but maybe it made them have second thoughts.

"When I was home I would put it in our tzedakah box," he said. "Joke's on them. They donated to a Jewish charity."[6]

Another young man, who asked to remain anonymous but said I could identify him as a Persian Jew in a chapter of AEPi, a Jewish fraternity, spoke of the stereotype but also said that he has come to think of "greedy Jew" or "cheap Jew" as a savvy person.[7]

Perhaps more significantly for the nation than the scourge of school-age penny throwers, there is the trope of Jews buying elections. If remarking to a schoolmate about how much Jews love money is a micro example of the old trope, then the macro example is House majority leader Kevin McCarthy tweeting, ahead of the 2018 midterms, "We cannot allow Soros, Steyer, and Bloomberg to BUY this election!"[8] or far-right talking head Tucker Carlson alleging that George Soros, through his donations to progressive district attorney candidates, was "hijacking democracy."[9]

George Soros, philanthropist, financier, and billionaire, is a major political donor, but to accuse him of hijacking democracy is to over-assign agency to him and strip it from voters and candidates. It also, once again, plays on the image of Jew as a perpetual outsider or for-eigner, trying to infiltrate the nation so as to subvert it.[10]

But this use of the trope of Jews and money does something else, too. It obscures the real, complicated, and fraught history American Jews have had with money on both ends of the spectrum: that of labor movements and that of finance.

There was, at the beginning of the twentieth century, a robust world of the Jewish Left. There was the legacy of the Bund, yes, but it was a living legacy: of organizing, protesting, and centering Jewish life around labor. The advent of the five-day workweek was, in fact, pushed by Jews in the United States. Jewish workers couldn't observe their Sabbath but also couldn't do shopping on Sundays, since that was off at the behest of their non-Jewish neighbors. In 1908 in New England, the (largely Jewish) managers and workers at one spinning mill came up with the solution: have both Saturdays and Sundays off.[11] The idea gained steam from there, with Jews and Christians alike working together to promote the two-day weekend. Predominantly Jewish labor unions in New York City pushed the five-day workweek as a priority.[12] Increasingly, employers, who noticed that the work got done in five days (with more contented workers, whatever their faith), signed on to the idea, too.

So, too, was there the belief held by some Jews from Russia and Eastern Europe that this was the more authentically Jewish way to be. As Yuri Slezkine recounts in *The Jewish Century*, "In the United States, which had no imminent perfection to offer, the memory of Russia—as the world of Pushkin and Populism—shaped the imagination of many first-generation immigrants." He points to a character in Abraham Cahan's *The Rise of David Levinsky*, Mr. Tevkin, who muses, "Russia is a better country than America, anyhow, even if she is oppressed by a Tsar. . . . There is too much materialism here, too much hurry and too much prose, and—yes, too much machinery. It's all very well to make shoes or bread by machinery, but alas! The things of the spirit, too, seem to be machine-made in America."[13]

But this workers' culture—this focus on labor, this memory of a revolutionary dream, this Yiddishkeit—eventually receded from the fore. History happened.

For one thing, the textile industry, which was heavily dominated by Jewish immigrants, turned out to lend itself well to upward mobility.

Jews went from being unskilled to increasingly skilled labor and from worker to business owner. The industry required workers to stay attuned to the world around them to keep up with changing trends. It also allowed those who managed to be titans of this particular industry to stay tied to New York, which itself allowed them to remain tied to the nation's financial capital.[14]

But all American Jews did not stay concentrated in urban centers and work-intensive industries. And, in fact, they did not stay in the garment industry, or most of them did not. By 1969, Jews, who were fewer than 3 percent of the US population, made up 27 percent of law faculties, 23 percent of medical faculties, and 22 percent of biochemistry professors. In 1970 at Yale, which had a quota on Jewish admissions until well into the twentieth century, 18 percent of professors were Jewish. As Slezkine puts it, Jews "stopped being exiled rebels in order to become salaried professors. A Russian-style prophetic intelligentsia had been transformed into a large contingent of rigorously trained intellectuals ('bourgeois experts') organized into professional corporations."[15]

Obviously, not all Jews went from making shirts to lecturing in universities. But America happened to them, which meant capitalism happened to them. And most Jews assimilated and acculturated. They moved out of the cities and into the suburbs. They stopped organizing their lives around social organizing.

"Of course, in the 1920s, socialism was very much part of the Workers Circle identity," Ann Toback, who was head of the Arbeter Ring, now known as the Workers Circle, told me over the phone. Its members were themselves tied to or related to those tied to the Bund. Yiddish was part of membership, yes, but so, too, was labor activism and union membership. These were part of their identities, and part of what grounded them in America.

By the 1950s and '60s, though, the first iteration of the Workers Circle—a fraternal benefits society founded by immigrants that served as a communal center for their activist journey into becoming Americans—was less relevant. There were far fewer Jewish immigrants to acclimate. It remained a very New York–based communal benefits organization (the communal benefits continued until the early 2000s) but was increasingly something of a relic. "They hadn't really been renewing themselves," Toback said.[16]

But the Socialist leanings of American Jews did not simply gradually dissipate. They were also actively snuffed out. From the 1920s on, there was a kind of war within American Jewish communities, Hasia Diner, director of the Goldstein-Goren Center for American Jewish History at New York University, told me. Because of both (1) their fear of antisemitism and the association in non-Jewish America's mind of Jews with Bolshevism and (2) their desire to assimilate and acculturate, Jewish institutional leaders wanted to ferret out Communism and Socialism from American Jewish life.

"They didn't like Communists, didn't like behavior within communal institutions, disagreed with the philosophy, and so on," Diner said.[17] In the 1950s, for example, the American Jewish Committee shared its files on suspected Jewish Communists with the House Committee on Un-American Activities.[18]

In more recent years, some American Jews—and in particular, young American Jews—have actively returned to Socialism, labor, and activism, and a Jewish identity centered around progressive organizing. But over the course of the twentieth century, the other side of the same coin developed: that of American Jews in the US financial system.

There have been Jews in American finance since before the wave of mass migration in the early twentieth century. In the mid-1800s, for example, Henry Lehman came to the United States from Bavaria. He was soon joined by his younger brothers, Emanuel and Mayer. In 1850, Lehman Brothers was born.

Though Lehman Brothers began as a dry goods store, they soon moved into other commodities—including, notably, cotton. Cotton was Alabama's main export crop, and so the Lehmans ended up accepting it in exchange for merchandise, which eventually led them into the cotton trading business. Henry Lehman died of yellow fever in 1855, but, in 1858, Emanuel Lehman was able to open an office in New York, the commodity trading center of the United States. Still, the brothers supported the Confederacy during the Civil War: Mayer Lehman provided financial support to Alabama's prisoners of war.

After the war, the Lehmans moved their headquarters up to New York. In 1870, they led the establishment of the New York Cotton Exchange.

And then, as the century turned, so, too, did the Lehmans' main business. The firm moved into railway bonds and then, finally, into the financial advisory business. In 1887, it became a member of the New York Stock Exchange. Roughly twenty years later, in 1906, they moved into investment banking , doing so with another firm, Goldman Sachs. Goldman Sachs was similarly founded by a German Jewish immigrant, Marcus Goldman, who came to America in 1848. He founded the firm along with his son-in-law, Samuel Sachs, who was born in Maryland to Jewish immigrants from Bavaria. (They did not have such good fortune forever. In the financial crisis of 2008, Lehman Brothers declared bankruptcy and Goldman Sachs would receive a $10 billion government bailout.)[19]

"The prominent role of German Jews in American finance was primarily a creation of the nineteenth century," Jerry Muller, author of *Capitalism and the Jews*, explained to me. Many of those banking houses remained active into the twentieth and even twenty-first centuries.[20]

There were other equally notorious examples. There was once a multinational investment bank by the name of Drexel Burnham

Lambert, born from a merger between the Drexel Firestone banking firm, based in Philadelphia, that could trace its history back to 1838, and Burnham & Company, a brokerage founded in 1935. Burnham & Company was founded by I. W. Burnham II, who wrote a letter to the *New York Times* in 1986 arguing that he felt there was great antisemitism on Wall Street before World War II but that it began to seriously decline in the 1960s. Wall Street, he wrote, was one of the country's "shining examples of integration at work."[21]

In 1971, a Jewish American by the name of Michael Milken took over the bond-trading department. Milken believed there was potential in so-called "junk bonds," or non–investment grade bonds given out by smaller companies or less fortunate established firms. By 1984, Drexel Burnham raised large amounts of capital floating new issues of junk bonds. This led to a mid- to late '80s "merger mania," in which Milken's clients and partners were able to oversee corporate mergers, acquisitions, and buyouts. Drexel Burnham became one of the country's most powerful financial firms.

But then, in 1986, a Milken client, Ivan Boesky, an American Jewish stock trader, was convicted of insider trading. He implicated both Milken and Drexel Burnham. In 1988, Drexel Burnham and Milken were both charged with securities fraud. Milken left the firm the next year, and, shortly thereafter, the firm went bankrupt. In 1990, Milken pleaded guilty to six counts of securities fraud and was sentenced to ten years in prison[22] (his sentence was reduced to time served in 1993, and he was pardoned by Donald Trump in 2020, citing the "incredible job" Milken had done contributing to cancer research following his release).[23]

In 1990, the *Washington Post* ran an article under the headline, "Undercurrent of Religious, Cultural Antagonism in Drexel's Rise, Fall." Counter to Burnham's 1986 letter to the *Times*, the *Post* piece presented a picture in which Wall Street was not particularly integrated, and in which there was animosity toward the Jews who worked on it. "Throughout the rise of the junk bond market, there was always a dark undercurrent of anti-Semitism at work: There is never any question that Milken and the highly visible junk bond raiders he financed were mostly Jewish," journalist David Warsh wrote. "Neither is there any doubt that their targets were usually companies run or owned by WASPs.

"The fact is that religious and cultural antagonisms have been an unmistakable part of the daily life of Wall Street for years. When Warren Buffett rescued Salomon Brothers from the clutches of the raider Ronald Perelman, some traders griped that an Omaha Episcopalian had rescued a 'Jew with a Christmas tree,'" Warsh went on.[24]

When James Stewart published his account of the whole sordid affair in a book called *Den of Thieves*, Milken's lawyer, Alan Dershowitz, charged that the book was a "vicious antisemitic diatribe," since four of the thieves in question were Jewish.[25]

Multiple things can be true. There could have been extra animus toward Drexel Burnham and Milken himself because of antisemitism. It could have even possibly brought greater attention to the firm and its misdeeds. But that doesn't change the fact that they did the deeds.

Given the origins of these famed firms, it is perhaps worth asking: Did Jews bring finance—with all the excesses and corruption it implies—to America? Or did America push Jewish immigrants into finance?

Unsurprisingly, the answer is a bit of both.

Jews in Europe worked in mercantile professions for centuries, in part because that was where they *could* work, and because Jews were not bound by religion against lending money. For hundreds of years, "court Jews," or Jews with connections to the most powerful families of Europe, financed Europe's aristocrats. The Rothschild banking house was established by one such "court Jew," Mayer Amschel Rothschild, and became an international juggernaut[26] (and a one-word wink at conspiracy theories that Jews controlled the world). And so, when Jews came to the United States, they brought that skill set and history with them. Still, it is not as though Jews came to the United States and invented finance.

There is a difference, Jerry Muller said, between areas where Jews were institutional innovators—that is, where they created something that wasn't there before—and areas where Jews happened to do very well. There were some industries, like Hollywood and the entertainment area, that were created in part by the contributions of American Jews. The film industry was created and shaped by Jewish studio executives, writers, and producers.

By contrast, upwardly socially economic professions like law, medicine, and finance were not invented in the United States by Jews. But Jewish immigrants in the nineteenth century were able to come over with financial experience and establish banking houses, and their children were able to follow in their footsteps.

Still, Muller allowed, it can be hard, sometimes, to differentiate between areas where Jews were innovators and areas where Jews were simply doing well. Capitalism and entrepreneurship, after all, are about finding niches and filling them. And in filling those niches, American Jews changed the very fabric of American enterprise, stepping, at times, on the fault lines of America itself.[27] Consider my great-grandfather, realizing that everyone wanted to buy ice.

Or consider, instead, the case of South of the Border.

South of the Border is a roadside attraction, so named because it is directly south of the border between North and South Carolina. It was founded by Alan Schafer in 1949 as a beer depot. The neighboring counties in North Carolina (north of the border, if you will) were dry counties—that is, they did not sell alcohol—leading South of the Border to become successful quite quickly.

In 1954, Schafer added motel rooms. He also started selling what the South of the Border website describes as "Mexican trinkets and numerous kitsch items imported from Mexico."

Schafer was not Mexican. He was an American Jew and a Southerner. But Schafer had gone to Mexico to make "import connections," where he met two young men whom he brought back to work at the motel as bellhops. On the South of the Border website, one can learn, "People started calling them Pedro and Pancho, and eventually just Pedro," though one cannot learn what their actual names were. The giant signs for South of the Border, featuring a racialized caricature of a Mexican man, were designed by Schafer.[28]

When accused of appropriating Mexican culture, Schafer told the *Washington Post* in 1979, he "plays on being Jewish in a small, Southern community. They usually come back with a sympathetic letter saying, 'Oh, you poor baby.' Once, a Mexican embassy guy wrote to a senator from New Mexico saying his embassy was hot, that we gave employers a bad image of Mexicans. I told the senator we had 100 good-paying jobs, above the minimum wage, with chances for advancement, and he should send some Mexicans down. I never heard from him again. They lost a chance to give jobs to 100 Mexicans."[29]

If Schafer's relationship with Mexican culture was exploitative, his relationship to Black Americans was more complicated. His grandson said that, as a child, he could remember the Ku Klux Klan burning crosses in their yard, though he wasn't sure if it was because they were Jewish or because they hired Black workers. And Schafer campaigned for civil rights supporters and worked to register Black voters.[30]

"I went out and registered every black citizen in the Little Rock precinct," Schafer said in that 1979 article. "Then I took control of that nucleus of 140 to 150 voters and I've had it ever since. With that black base, I took over the county machine. The Ku Klux Klan used to follow the trucks of my beer-distributing company around. I was a pariah in the white community."[31] But he also added plainly racist attractions like "Confederateland, USA" and Pedro's Plantation, where tourists could pick cotton.[32]

Who was Alan Schafer: Jewish Southerner? Cultural appropriator? Racist? Civil rights champion? Klan adversary? Exploiter of Black Americans for his own political and pecuniary purposes? A capitalist, who, as Muller described, found niches and filled them?

Or all of the above?

The part of the story of American Jews and labor and finance that traditional narratives most often leave out is that not all Jews are upper or even middle class. There are American Jews who fit comfortably neither in the story of the acculturation of the Jewish Left nor in the tale of American finance. There are Jews who still, today, live in poverty.

This is not a new phenomenon. In her 1974 book *Poor Jews*, Ann G. Wolfe included a chapter titled, "The Invisible Jewish Poor," a presentation she had delivered a few years earlier. In it, she describes that, though only several years prior, Americans had learned that thirty million lived below the poverty line, American Jews did not appear to believe that that statistic had much to do with them. The Jewish poor, she wrote, had problems like "poor housing, inadequate medical care, neighborhoods that are undesirable in terms of emotional and physical security and outside the Jewish cultural mainstream." But though they were out of the mainstream, and though their stories are not the stories many Jews like to tell themselves about what it is to be Jewish, they still exist. "How are priorities set?" Wolfe asked. "Where is the power? Does the Jewish community need to re-order its domestic priorities?"[33]

Almost forty years later, a report came out that painted a startlingly similar picture to the one that appears in Wolfe's writing.

In 2019, a report by Jonathan Hornstein for the Harry and Jeanette

Weinberg Foundation found that roughly a fifth of Jewish households earned less than $30,000 a year. It is true that 20 percent is not
50 or 75 percent, but it's also not nothing. And Jewish poverty, and
the reality of Jewish lives that are in poverty, is obscured not only by
stereotypes that Jews control or are disproportionately represented
in finance but also by the story some American Jews tell that Jews
rose through the ranks of American meritocracy, working hard and
pulling themselves up by their own bootstraps with intelligence
and ingenuity. In truth, the system worked for some American Jews,
and some American Jews were able to make it work even better for
them. But for thousands, the system did not, and they were not able
to improve it.

For all the association of Jews with Wall Street and finance, today New York City and the surrounding area had the largest number
of low-income Jewish households in the United States. Jewish poverty was particularly concentrated in, among others, Haredi Jews;
less educated Jews; older Jews; and Russian-speaking Jews. These are
not, incidentally, necessarily distinct communities. For example, the
Pew Research Center found that 43 percent of traditional Orthodox
Jews had a household income below $50,000, compared to 32 per-
cent of Jews overall. But Haredi Jews were also less likely to have
as much secular education; only 25 percent had a bachelor's degree
or higher, compared to 58 percent of overall Jews. And some of
the older Jews were also Russian-speaking Jews; 72 percent of this
group—that is, older Russian-speaking Jews, some of whom had no
history of work in the United States and no access to social security or
benefits—lived in poverty.[34] (Incidentally, according to Wolfe, elderly
Jews were more likely to be poor back in 1971, too, left with lack of
appropriate housing and clinging to areas that were no longer "Jewish." The individuals of whom the elderly population is composed
has changed; that elderly Jews are more likely to be impoverished
has not.)[35]

Perhaps most notably, given the space that Holocaust memory takes up in American Jewish life, about one third of Holocaust survivors in the United States live in poverty, according to the Blue Card foundation (there is, of course, also overlap here with the older Russian-speaking Jews). The Blue Card helps survivors, as do other groups, like KAVOD (or *dignity*), which works to provide survivors with emergency needs. KAVOD was founded by John and Amy Israel Pregulman. In 2019, KAVOD and the Seed the Dream Foundation created the KAVOD Survivors of the Holocaust Emergency Fund. In mere weeks, they had eighteen thousand requests.[36]

All of which is to say that Jewish poverty and Jewish wealth are entangled with another political reality of Jewish America: Jewish philanthropy.

There is, in this country, a long history of more established Jews raising money for other Jews. Acculturated Jews provided assistance—and unsolicited advice—to their eastern European immigrant coreligionists. The Workers Circle, back when it was Workmen's Circle, may have grounded itself in Yiddishkeit and labor rights, but it was also a kind of philanthropy.

But then came the 1930s. The Great Depression, which shook Jewish philanthropic and institutional life, played a role in shaping Jewish philanthropy. Increasing American Jewish acculturation and affluence also helped. But the Holocaust was a factor, too.

Charities and philanthropic entities began to save up, to collect, to conserve. It was part of the "way that American Jews processed what the hell this meant, what this existential crisis demanded of them," Lila Corwin Berman, author of *The American Jewish Philanthropic Complex: The History of a Multibillion-Dollar Institution*, told me. American Jews became consumed with the idea of survival in a way that they hadn't before. And with "different tools that would make survival more likely." Jewish philanthropy was one.

"I'll never forget talking to this guy from a midwestern federation," she told me. "He said: 'We may not have any Jews who live in the city, but we will always have these endowments.'"

There are 146 Jewish federations in North America and 300 "smaller communities." Their stated mission is "to protect and enhance the well-being of Jews worldwide through meaningful contributions to community, Israel and civil society."[37]

"I just can't imagine the kind of transformations that happened without thinking about how central that constellation really was," Corwin Berman said.

Other evolutions happened, too.

For one thing, over the course of the twentieth century, America's tax structure and systems changed dramatically.

As Corwin Berman, who is a history professor at Temple University, documents in her book, while this wasn't only because of the work of American Jews, individual American Jews did play their part in shaping this part of history. One such individual was Norman Sugarman. Sugarman, a midwestern tax lawyer, advised the Jewish Federation of Cleveland, among other entities, as they grew their endowments. Along the way, he created the concept of a "donor advisory fund," which lets people put their money under public charities' legal stewardship, and then pushed for donor advisory funds (today, donor-advised funds) to be recognized as legally legitimate. It worked.[38]

That isn't to say that Sugarman was wholly responsible. "I tend to think that the structure of American law and policy was necessary for anyone like Sugarman to do what he did," Corwin Berman said. She noted that, for example, over the past several decades, tax rates changed, too. After World War II, the highest marginal tax rate was over 90 percent. In 2020, it was 37 percent. "There has just been a total, total shift in terms of thinking about private property and individual rights over private property," Corwin Berman said.

Still, Sugarman "was also someone who as an individual was really smart, creative, and had a particular interest in accruing capital."[39]

And it is at least in part thanks to him that, today, American Jewish giving is estimated at $6 billion a year. In 2013, Jewish federations had an estimated $16 billion in assets[40] (and that's even with Jewish federations' lost primacy in American Jewish philanthropic space).[41]

As the endowments expanded, the politics of mainstream Jewish philanthropic institutions shifted, too.

"When you have an organization, and part of its goal is to hold wealth and have some kind of distribution policy that is still tied to the continued accumulation or growth of that wealth—it's a conserving tendency," Corwin Berman told me. There was, baked into the foundation of Jewish philanthropy, this conservative element.

That conservative tendency or element was not necessarily bound to evolve into more conservative politics. But history happened. Or, as Corwin Berman put it to me, "very specific historical circumstances of American and American Jewish life did make those kinds of groups in general invested in a conserving kind of ideal of what Jewish identity was about."[42]

One of those ideals was endogamy, or the tradition of only marrying within a particular clan or community. Part of this came out of the Holocaust and the decimation of Europe's Jewish population. But another part of it was carefully curated.

There exists, within the Jewish communal world, a debate between assimilationists and transformationalists. The former, broadly speaking, believe that if Jews intermarry, Jews will die out. The latter believe that while social change will shape what Judaism and Jewish life looks like, that, in its own way, is continuity, because Judaism and Jewish life have been changing and evolving for as long as there has been Judaism.

The stress on the Jewish family, which had long been an element in Jewish communal life, also coincided neatly with where mainstream

America was in the postwar Cold War period, in which the family unit was a bulwark against Communism. Jewish philanthropic organizations paid Jewish social scientists to go out and collect data on endogamy and Jewish tradition and practice—and then went out and used the data collected to push a message against intermarriage. If, for example, a survey collects information that says that Jews who marry other Jews are more likely to be synagogue members, or light Shabbat candles, or socialize mostly or exclusively with other Jews, that is a quantitative assessment. But it isn't only quantitative, as Keren McGinity, who teaches in the American Studies program at Brandeis, explained to me. "It really looks at Jewish identity and community and life through a very, very narrow lens."[43]

Nevertheless, these studies, funded by Jewish philanthropic wealth, reinforced a fear factor, and the idea that people who didn't marry Jews didn't care about being Jewish or participating in "the Jewish community." The very idea that there is a singular Jewish community is itself a contested concept but one pushed by the leaders of established, mainstream Jewish institutions.

The first National Jewish Population Survey was published in 1971; later iterations came in 1981 and 1990. And it was presented, as Kate Rosenblatt, assistant professor of religion at Emory University, put it to me, as being beyond rebuke. "All of this stuff has the imprint in academia on it: 'These are *numbers*, you can't argue with numbers.'

"The data is always a reflection of who's asking the questions and what kinds of questions they're asking," she said.[44] And the questions were being asked and prepared by men, who, incidentally, put the blame on Jewish women.

And when, in the 1990s, Jewish communal leaders—people on the ground trying to create some semblance of Jewish life in contemporary America—suggested that perhaps they try to make intermarried couples and their children feel more welcome, and that it was their own behavior that could change, these social scientists, backed by Jewish philanthropy, tut-tutted. These communal leaders were proposing moving Jewish resources from the *core* to the *periphery*. These intermarried Jews, after all, couldn't possibly care as much about being Jewish.[45] As Alan Dershowitz wrote in his 1997 book *The Vanishing American Jew*, "A decision by a young Jewish man or woman to marry a non-Jew is generally a reflection of a well-established reality that their Jewishness is not all that central to their identity."[46]

This attitude did not stay in the 1990s. In March 2021, Jewish philanthropist Matthew Bronfman, newly positioned as the chair of Hillel's Board of Governors, was asked what keeps him up at night. "Intermarriage," he said.[47]

I read this interview as a Jewish woman who had married a non-Jewish man a few months prior. I read this and thought of my husband, who joined a synagogue with me; lights Shabbat candles every week with me; hosts Passover seders with me; buys apples and honey for Rosh Hashanah; has his own menorah; watches movies about Jewish history and culture with me; goes to Jewish museums and memorials around the world with me; and agreed while we were still dating to raise our children Jewish. I thought of those hypothetical future children, whose last name will be "Bhatiya," and wondered what Bronfman would say about them.

And I wondered if he understood what it felt like to hear from people who are held up as Jewish leaders that the big threat to Jewishness is you, a person who is so proud to be Jewish and who happens to love someone who is not. I wondered what American Jewish communal life would look like if people like Bronfman thought less about whom I was procreating with and more about what it means to be Jewish, and how to create inclusive spaces in which people, regardless of whom they marry, could explore Jewishness for themselves without feeling judged. I wondered what message Bronfman thought he was sending to Jews who marry non-Jews.

And then I realized that I already had my answer. We were on the periphery. He was the core. He didn't care.

In that same interview, Bronfman says that there is another thing that keeps him up at night, which "plays a role in intermarriage, is what I view, and others view, as a growing disconnect between the next generation and Israel. The view that Israel is the oppressor. Goliath, not David. It leads them to not lead a Jewish life, and not necessarily want to marry Jewish, and that's why I think a trip to Israel—Birthright—is critically important."

If Jewish continuity and survival within America were one pillar of Jewish philanthropy in the twentieth century, Israel was another.

There are some who might argue that Israel was always bound to be a conservative cause. These people would also argue that Zionism—the idea of a Jewish homeland—was, from the very beginning, inevitably going to lead to Israel in the 2020s, a country regularly admonished by human rights groups. In this reading of history, Israel is first and foremost a settler colonial project, preordained to be a right-wing cause.

There are others who would note that Israel was initially home to Labor Zionism and Socialist infrastructure, and that to say that it was always bound to be right-wing is to acquit generations of political leaders and their attitudes and policies toward Palestinians.

But regardless of how one feels about Israel's founding and where Zionism falls on the ideological spectrum, one is left with the evolution of Israeli politics. Millions of people have been under occupation since 1967 (and people within Israel proper were denied full rights since before that). A two-state solution has, for decades, been held up as a solution for peace, but it is a solution that has never arrived, leading the status quo to calcify.

There are moments to which one could point for why this is, or why American Jews have become more critical of Israel. In 1995,

Yitzhak Rabin, prime minister of Israel and member of the Labor Party, was assassinated. Rabin was hardly a perfect progressive figure. As chief of staff of the Israel Defense Forces, it was he who led the IDF to victory in the 1967 war. As defense minister in 1987, he suppressed the First Intifada, an uprising of Palestinians. But he also came to believe that Palestinian resentments could not be pushed away with war and weapons and required a political solution. In 1993, he signed the Oslo Accords, in which Israel and the Palestinian Liberation Organization recognized each other and promised peaceful negotiations.[48] In July 1995, Benjamin Netanyahu, then simply leader of the Likud party, led a mock funeral at an anti-Rabin rally complete with coffin and noose.[49] Four months later, Rabin was assassinated. Netanyahu began his first stint as prime minister the following year. One could argue that the political Left in Israel has been slowly weakening ever since, and note that, two decades later, coverage of Israel's Left was dedicated to its decline and collapse. Or one could instead point to the Second Intifada, a Palestinian uprising that came from the failure of the peace process in the first decade of the 2000s and the violence that ensued, and to the fact that Israelis lived through it and American Jews did not and so have a different approach to politics and security,[50] or to any number of other things.

Regardless of which points one thinks were most important to get from 1948 to today, the reality remains that, in 2016, according to a Pew study, a full 48 percent of Israeli Jews said Arabs should be "transferred or expelled" from Israel.[51] In 2019, the Jewish Telegraphic Agency reported that young Israeli Jews—who, the publication noted, had never known a real peace process—were more likely to be right-wing than their parents,[52] which is to say that there is arguably less room, even on the relative left of the political spectrum, for progressive policy with respect to Palestinians.

This has meant that, increasingly, Jewish organizations that defend the Israeli government's behavior are defending right-wing political preferences. (There are Israeli Jews—and indeed American Jews—who argue that judging what Israel does from the United States is wrong, or at least unfair; that to sit comfortably in Washington or New York and judge the Israeli military and government is hypocritical and cheap. Whether or not one agrees with that, the reality remains that American progressive movements and champions of Israel are moving further apart.)

Relatedly, arguably the most prominent US donor to the cause of promoting love of Israel and Israeli culture in the United States among American Jews also happened to be a major Republican donor. Sheldon Adelson, the business magnate, did not found the Taglit-Birthright program, which provides a free trip to Israel for American Jews ages eighteen to thirty-two (it was expanded from eighteen to twenty-six)—it was cofounded by Charles Bronfman and Michael Steinhardt, the latter of whom had to return $70 million in stolen relics in 2021 and was accused by several women at nonprofits he supported of sexual harassment in 2019, among other donors and organizations.[53] But Adelson did pledge $25 million a year in 2007.[54] In 2018, to celebrate seventy years of Israeli independence, Adelson and his wife, Miriam, pledged $70 million.[55] In a 2013 speech, he called on the United States to drop a bomb on Iran.[56] (During the Obama years, Adelson also reportedly offered to pay for the Iron Dome, Israel's missile defense system; Obama reportedly declined.)[57] He funded numerous pro-Israel groups in Washington, DC, at various points of his life.[58]

The Adelsons also donated hundreds of millions of dollars to the Republican Party. Sheldon Adelson endorsed Trump in a 2016 *Washington Post* column[59] and gave him $25 million in support in the 2016 presidential election[60] (he threw in $5 million more for Trump's inauguration).[61] Sheldon Adelson was reportedly one of the driving forces in Trump's decision to move the US embassy from Tel Aviv to Jerusalem.[62] Miriam Adelson, to whom Trump awarded a presidential medal of freedom, said that she hoped that, just as there is a book of Esther, so there would one day be a book of Trump.[63]

All of which is to say that, while Israel still has bipartisan support in the United States—though it has increasingly been claimed by Republicans and questioned by Democrats—it has in many ways become increasingly conservative, leading those who unfailingly support it to defend conservative government policies. It is also to say that, while most American Jews vote Democrat, the reality is that some of the most prominent people who are pouring their money into Israel's promotion in the United States are also simultaneously supporting some of the most openly ethno-nationalist politicians in contemporary American history.

A particularly notable exception in the world of Jewish philanthropists is George Soros, the Hungarian-born billionaire whose Open Society Foundations focuses on creating societies in which everyone has an equitable chance at participation. Though he ventured into philanthropy in 1979, Soros did not give money to a Jewish cause until the early 1990s, and even when he did—to a Jewish organization in Bosnia run by a man named Jakob Finci—he made his recipient assure him that the money would go to all people, not just Jews.[64]

Writing in the Jewish magazine Tablet in 2020, the author and military strategist Edward Luttwak recalled telling Soros that "he stop funding Hungarian causes and instead give his spare cash to Israel and its then great number of new immigrants from Russia. I did not claim that the Hungarians were especially undeserving—I merely pointed out that he, Soros, would inevitably unleash anti-Semitic reactions by donating money to Hungarian causes."

Hungarians, he wrote, "would think it impossibly naive to believe that Soros was motivated by generosity alone; no, there had to be a sinister, Judaic plot, even if non-Jews could not be expected to identify it exactly by unraveling its Talmudic complexities."

He concluded his piece by stating, "I take no satisfaction from having been proved 100% right and Soros 100% wrong."[65] And, in one sense, he was right: Conspiracy theories about Soros and his nefarious attempts to bring migrants into Hungary have become a mainstay of Hungarian politics. In the United States, Soros has been accused of everything from bringing in migrant caravans to smuggling drugs to hijacking democracy.

Luttwak's piece did not mention, however, that Benjamin Netanyahu, who was prime minister of Israel for twelve years, and his son Yair themselves pushed antisemitic conspiracies about Soros; or that Arthur Finkelstein, the political operative who helped Hungarian prime minister Viktor Orbán take Soros conspiracy theories mainstream, was introduced to Orbán by Netanyahu; or that Hungarians would hold Soros's relationship to Israel as proof that he wasn't a real Jew.[66]

But, then, that would have defeated the purpose: Luttwak concluded his article by asking Soros to consider whether it is not too late to divert his giving to Israel.[67]

One interviewee, who asked to remain anonymous, began our conversation by telling me, "I feel like a Bad Jew all the time." I asked him why.

My interviewee had grown up in a modern Orthodox family that was dedicated to Orthodoxy but also to Jewish civic engagement. And my interviewee set out to live a professional life in Jewish institutional spaces, funded by Jewish philanthropy.

"The perverse influence of big money in Jewish communal life is a constant influence," he said.

My interviewee was a Wexner fellow, meaning he received money from the Wexner Foundation, which invested in him as a future leader among Jewish professionals. But it should have been called something other than the Wexner fellowship, he said. It was named for its funder, Les Wexner, the billionaire behind Victoria's Secret.

Since Wexner was a close associate of Jeffrey Epstein, the fellowship is now tarnished, or so many of the fellows feel, according to my interviewee. Perhaps, had the name been something else, it "could have insulated the fellowship from his incredibly weird relationship with the most prominent pedophile in history."

But the issue is bigger than Wexner, he said.

Within philanthropic organizations, "the governing structure is such that the people who give are the ones who set the agenda, and the people who run it are paid employees.

"It's almost like, in all of the Jewish communal spaces, wealthy people parlay [their wealth into] claims of representing the Jewish community," which they then use "to pursue whatever bee is in their bonnet at any given time." And what Jewish community were they even representing? "The American Jewish community is a myth," he said. There are communities. But there's no one united bloc of people. Even if there had been, they didn't exactly come together to elect these wealthy individuals to say what matters to us. But here they are, doing exactly that.

My interviewee couldn't stand it. And so he walked away from it.

But he felt guilt over that—over walking away from a field because he couldn't stand the power that money had over it.

"What I feel a lot—I guess I was groomed to work in American Jewish life." There had been "so much invested" in him, he said. There was his family, that foundation and fellowship, other scholarships.

"I know a lot of rabbis, I know a lot of people who work as Jewish professionals or educators . . . maybe I should have been happy," he said.

"I wasn't. I don't know if that makes me a Bad Jew."[68]

I wanted to tell him that I disagreed with that. That I don't think that pushing back against something that you think is wrong, even if it's being done by people who are very powerful Jews, and even if it means stepping away from Jewish institutions, makes a person a Bad

Jew. I wanted to say that what he was doing was questioning and challenging, which are two of the Jewish traditions that I care about the
most. I could have also said that he wasn't alone in feeling this way. Al
Rosenberg, chief strategy officer at OneTable, a nonprofit that is dedicated to helping people build their own Shabbat dinner practice,
told me, "I have a lot of hopes and dreams for the future of Jewish
community." And one of those hopes is "decentralized funding."

"If the funding of a community can be aimed at making ritual
accessible to its community members, I think Judaism would be celebrated more. People would be more excited to be a part of it," Rosenberg told me. But right now, they said, there is a gap between many
institutions—and their money—and the people those institutions ostensibly serve.[69]

Even those associated with establishment Jewish institutions were
not without criticism of Jewish philanthropy. A man who asked to
be referred to as Adam has long been involved with Jewish philanthropy and has had leadership roles in Jewish organizations. Still, he
said, he and his wife are "not big federation people." Federations, he
said, tend to believe that "God is the right arm of the federation, and
it's all about the money."[70] Meanwhile, I spoke to one person who
was critical of federations but noted that federations at least offered
an element of democracy that they worry is slowly leaving Jewish
philanthropic life.[71]

I could have told all of this to my anonymous interviewee. That
he wasn't making any of this up. That this is a real problem, and
that people are hungry for more from their Jewish institutions, and
that those institutions can't hear them because they are too far removed by all that money.

But I didn't. Instead, I just thanked him for his time.

Where does all of this leave us with respect to stereotypes of Jews
and money?

It leaves us with the reality that many Jews are and have been involved in finance and Jewish philanthropy, just as many Jews are and have been involved with various labor and workers' rights movements.

Americans, Jews, and American Jews shouldn't—and likely couldn't even if they wanted to—stop talking about various Jewish relationships to money. But how we talk about it matters.

When Mike Bloomberg, billionaire and former mayor of New York, ran for the 2020 Democratic presidential nomination, he poured over a billion of his own money, made via his financial information (and software and media) firm, into the race, for himself. When Republican Texas senator Ted Cruz tweeted of Bloomberg, "It's almost as if he controls the media"—not one particular media outlet (which would have been true, since Bloomberg does own *Bloomberg*), but all of the media—he crossed the line from considered critique into assignment of agency.[72]

But to just observe that Bloomberg was spending hundreds of millions of his own money on a presidential campaign cannot be simply written off as antisemitism. To go a step further and say, for example, that the very fact that one person could make and be allowed to keep enough money that they could fund their own campaign for president was a searing indictment on the threat of oligarchy to democracy isn't antisemitism, either.

Writing in *Politico*, journalist Jack Shafer, who is Jewish, wrote, "Can you locate a significant Democratic Party constituency that has not become beholden to Bloomberg's money?," and then very carefully traced and tracked where in the political arena Bloomberg had and had not spent money. "The American people are sick and tired of billionaires buying elections" might have been an antisemitic line, had it been said about the influence of Jewish individuals on politics, but it wasn't; it was uttered by Bernie Sanders, who, in addition to also being Jewish (and also running for president that cycle), was explaining why he was confident that he, Sanders, could beat Bloomberg. There is plenty of rhetoric about Jews and politics and finance that is antisemitic, but noting that a Jewish person has a lot of money and was using that money to try to further a political campaign is not antisemitic or evident of hatred against Jewish people. It does no great service to Jewish people to pretend we are above rebuke.[73]

Another, darker example is the case of Bernie Madoff, an American financier who ran the largest Ponzi scheme in history—many of the victims of which were other wealthy Jews. A former Nasdaq chairman, Madoff, for decades, cultivated a reputation of a financial wizard who got his clients predictably safe returns on their investments. In 2008, though, it was revealed that it was all smoke and mirrors: Madoff was actually using new investors' money to pay off old investors' returns. Phony investor account statements at the time of his arrest claimed to hold $65 billion.

Madoff did not only steal from wealthy Jews, but also from Jewish charities and institutions. The American Jewish Congress lost millions of dollars. Madoff defrauded Elie Wiesel, Nobel Prize winner and Holocaust survivor, who said in 2009 that he, Wiesel, had fallen for "a myth that he created around him that everything was so special, so unique, that it had to be secret." Madoff cultivated an air of exclusivity, a network into which knowing Jews (or rather, Jews who thought they were knowing Jews) brought other Jews.[74]

My mom told me that, when she hears the phrase "Bad Jew," a picture of Madoff pops into her mind.

"He did everything you're not supposed to do as a Jew," she said. "He stole from people, he ruined people's lives. He actually stole from his fellow Jews. He went back to his own and stole from them. He took advantage of a connection that people felt with him. . . . I think that's the worst kind of Jew there is."[75]

"I went to Brandeis," Shelby Magid, an associate director at the Atlantic Council, a think tank in Washington, DC, told me. "He screwed over so many of our big donors."

She still didn't like the term "Bad Jew," she said. "I would maybe reserve it for cases like that, where you're doing wrong to your own community.

"He was a bad one," she said. Then she added, "But he was a bad *person*."[76]

There is some discomfort in talking about Madoff—his lack of scruples, his money and obvious desire for more of it, and his disproportionately Jewish victims—because it so neatly plays to the stereotype of the cunning, money-hungry Jew.

But not talking about what Madoff did doesn't change the fact that he did it. As with Bloomberg, and as with Drexel and Milken, there is, of course, reason to be careful and considered in discussing and debating what Madoff did, and in how many Jews he robbed. But to refuse to discuss it is to refuse reality. There are Jews who are involved in capitalism and its excesses; that doesn't mean that we should shy away from critiquing capitalism, any more than the existence of Jewish neoconservatives means we shouldn't examine American imperialism, or that the fact that Jews performed whiteness means we can't challenge white supremacy. We must examine these phenomena, and even certain Jews' roles in them; we also shouldn't blame Jews qua Jews for them.

To refuse to discuss it for fear of worsening hateful stereotypes does something else, too: it pretends that the stereotypes exist because of what Jews do or do not do, and not because they are serving some other, hateful purpose.

"Those tropes and fantasies reflect much, much more about the people who are spinning them . . . than they do about Jews themselves," Corwin Berman said. "In fact, silencing those conversations has much more danger than actually having those conversations."[77]

• • •

The story of American Jewishness and American Jews and labor and wealth and poverty is one of ebbs and flows. It is one that evolves from the early twentieth century, with its new Jewish immigrants and their social origins, through a century that saw tax policy and government attitude toward philanthropy and private wealth change and saw finance and industry and entrepreneurship expand—not entirely but in part because of certain Jewish actors.

This story is not linear, and it is not without competing narratives. But it is complicated still further when one considers that, while the early twentieth century was the high-water mark for American Jewish immigration, there were still Jews who came as immigrants to America after that. They would put their own mark—or, rather, their own conflicting, competing marks—on American Jewish history.

Chapter 7

Refugee Jews

OLYA WAS BORN IN the Soviet Union. Her father had grown up in a fairly religious household; her mother, in a secular one. Her mother's first Passover seder was at Olya's father's parents' home. For her father, Judaism was a religion. For her mother, in the Soviet Union, it was "a form of protest."

"For both of them," Olya told me, "it was something that caused discrimination and made their lives more difficult."

Olya, who asked to be identified by only her first name, and her parents were refuseniks—people, primarily Soviet Jews, who were refused the right to emigrate. Her family finally, with help from the American Jewish community, which lobbied on behalf of their Soviet coreligionists, came to the United States when she was six years old.[1]

Olya and her family's story stands in contrast to that of many of the American Jews whose families came over in the late nineteenth and early twentieth centuries, but Olya's family was hardly alone. In September 1989, the *New York Times* reported on the arrival of the largest group of Jewish refugees to come to the United States in a single day since World War II (on that day, 1,356 Soviet Jews reached the United States, joined by roughly 250 Pentecostal Christians). They arrived in time for Rosh Hashanah, the Jewish new year.[2]

The passing of legislation in the 1920s that severely restricted Jewish immigration was significant in part because, without an influx of Jews primarily from central, eastern, and southeastern Europe, assimilation and acculturation intensified, as did the contestation over what it meant to be an American Jew.

But though there were fewer Jewish immigrants to arrive after 1924, there were indeed still some Jewish populations who found their way to the United States. They came from different countries and contexts and had been forged by different political circumstances, but they shaped, and still shape, what it means to be Jewish in America differently than did many American Jews whose families came to the United States earlier, too.

This chapter looks at two such groups: Soviet and Persian Jews, thousands of whom fled to the United States in the 1970s, during the Islamic Revolution and the end of the Persian monarchy.[3] Though there are many differences between them (and, indeed, there are many differences *within* both groups), both serve as a reminder that there is more than one American Jewish immigration—and assimilation—story.

There was a moment, at the beginning of the early twentieth century, when Jews let themselves believe that the Soviet Union would be their true home. That it was there that Socialism would bring about the conditions for Jewish intellectualism and culture to flourish. That it was in the Soviet Union that Jews could be their most authentic selves, a condition somehow always just out of reality's reach.[4]

It was a fleeting moment. The Soviet Union would not provide a haven for Jews.

Despite the trope that Jews were all secret Communists, in the actual Soviet Union, Communism, as it developed, was hostile to Jewish practice. In the early days of the Soviet Union, various minority nationalities were actually encouraged to thrive, and so Joseph Stalin, as the first Soviet government's commissar of nationalities, encouraged Yiddish culture and education, and even worked on the establishment of an autonomous Jewish region, Birobidzhan, out in Russia's far east. It was Stalin who gave the permit for Moscow's Habimah, a Hebrew theater.

But Stalin eventually reversed the nationalities' policy of his predecessor, Vladimir Lenin. He liquidated Yiddish publications, theaters, and schools; by the end of the 1930s, only a handful remained. He may have supported the establishment of Israel, but, during World War II, he resisted a plan to allow Jewish evacuees to return to Crimea. A "Jewish Crimea," he believed, would pose a security risk to the Soviet Union. And from 1948 until his death in 1953, Stalin was virulently opposed to Jewishness and Jews themselves. Even the last Jewish institutions were liquidated. Jewish writers and artists, intellectuals and professionals were arrested en masse. His campaign against "rootless cosmopolitans" obviously (though not exclusively) targeted Jews. Twenty-six leaders of Jewish life were secretly tried and, on August 12, 1952, executed, accused of Jewish nationalism (and of trying to detach Crimea from the Soviet Union). Six of the nine doctors Stalin accused of trying to poison Soviet leadership in the so-called Doctors' Plot were Jewish.[5] His death was the only thing that halted their trial and, many believe, a year of disaster for Soviet Jews. (Incidentally, though he died on March 5, Stalin fell ill on March 1, 1953—Purim, a Jewish holiday that marks, among other things, the downfall of a man who plotted Jewish destruction.) The period after Stalin's death was marked by his successor, Nikita Khrushchev, admitting Stalin's excesses, and even introducing some small room for creative expression and freedom in the period known as the Thaw—which lasted only a few years before leadership felt that even these few relative relaxations were threatening to the regime.[6]

Various Soviet Jews responded to all of this—the repression, the threats, the choking off of their own culture and religion, the introduction and cessation of freedoms—differently. Some tried to simply go along to get along.

A very small number pushed back not for the rights of Jews specifically, but for rule of law. Led by Aleksandr Esenin-Vol'pin, an eccentric mathematician and the illegitimate son of famed poet Sergei Esenin, this small group of individuals fought—and in many cases spent time in prison—for the right of every person to be accorded his legal rights, and for everyone, even those in power, to be accountable to the Soviet constitution. Esenin-Vol'pin was Jewish.[7] So were others in the legal dissident cohort, like Aleksandr Ginzburg.[8] Jews were, in fact, disproportionately represented in the rights-based dissident cohort. Being Jewish was not at the center of their work. For one thing, those who wanted to focus on their rights as Jews, as opposed to citizens of the Soviet Union, found that they did not have a place in this movement. For another, the driving principle was not a celebration of Jewishness or Judaism but individual integrity in an oppressive system. When I interviewed her in 2013, Aleksandr Ginzburg's widow, Arina Ginzburg, likened her group of dissidents to a warm orb within the cold mass of the Soviet system.[9] The point was to keep that warmth and connectedness and individuality among others who appreciated the need for it. Still, though the legal dissidents may not have centered their Jewishness, it is perhaps not coincidental that many of the people in this movement that said everyone was equal before the law were Jewish, given the unequal treatment of Jews in the Soviet Union.

But there were also refuseniks—Soviet Jews who rallied for their right specifically as Soviet Jews to leave the country. There was, of course, an element of the fight for universal rights in this movement, too: the right to freely emigrate. But unlike the legal dissidents, the refuseniks' was an inherently identity-based campaign. There was crossover between the two—or, rather, there were rights-based dissidents who became refuseniks, arguably most famously including Natan Sharansky.

It was the refuseniks who captured the imagination, sympathy, and energy of American Jewish institutions and community leaders. Some have argued that it was this moment in American Jewish history when American Jews learned to lobby.[10] This is not, strictly speaking, true. American Jewish institutions and individuals lobbied over immigration—against changing the legal racial status of Jewish immigrants, against immigration restrictions, and for immigration reform—and for creation of the state of Israel. Still, it is true that the American Jews who lobbied for Soviet Jews were stepping into their own power as Jews. If those who had lobbied for Jewish immigrants wanted Jews not to be seen differently, those who lobbied for Soviet Jews were comfortable distinguishing themselves and highlighting the distinct plight of Jews abroad.

American Jews clearly felt empowered, in the late twentieth century, to lobby for Jews as Jews in a way that they had not earlier on in American history. This was for a combination of reasons. For one thing, American Jews as a group were more established, assimilated, and acculturated; their status as Americans was less precarious than it had been at the beginning of the century. For another, as Marc Dollinger wrote in *Black Power, Jewish Politics*, the Black Power movement and its affirmation that a people could assert itself as a people served as a kind of model for American Jews. Some point to the late 1960s and the 1970s as a breaking point between Black Americans and American Jews in the civil rights movement, and to some extent this is true. But it is also true that, at this time, American Jews learned from Black Americans and picked up what were considered explicitly—even nationalistically—Jewish issues to champion.

All of this was resented, at least at times, by some of the legal dissidents, though they did consistently defend the refuseniks' right to leave the country; in her memoir, *The Thaw Generation*, Ludmilla Alexeyeva, a giant of the universal rights–based dissident movement, wrote that the universal rights–based dissidents were in a kind of competition with the refuseniks for Western attention. She felt the refuseniks garnered more headlines and support because their struggle for emigration was a simpler narrative for outsiders to understand and get behind than was universal human rights and rule of law.[11] In his memoirs, Andrei Sakharov, one of the most famous rights-based dissidents, allowed that the attempted hijacking of a plane in Leningrad in 1970 by a group of refuseniks was not wholly a human rights affair.[12]

Still, the attention and support garnered from American Jews was not for naught. Though originally President Richard Nixon's secretary of state, Henry Kissinger, despite himself having been a Jewish refugee from Nazi Germany, did not want to deal with the issue. The Nixon administration was pursuing détente with the Soviet Union, and what were some Jews stuck in the Soviet Union compared to détente in the mind of the man who believed himself a master of realpolitik?

"Let's face it," Kissinger said in 1973. "The emigration of Jews from the Soviet Union is not an objective of American foreign policy. And if they put Jews into gas chambers in the Soviet Union, it is not an American concern. It may be a humanitarian concern."[13]

But American Jewish activists forced the issue, finding a partner in Senator Henry Jackson. The Jackson-Vanik Amendment, a part of the US Trade Act of 1974, threatened to impact trade with non-market economies that limited Jewish emigration.[14]

In truth, the legislation did not spark a massive rush out of the Soviet Union, and there are those who doubt that it should be considered a success. Nevertheless, in time, Soviet leadership changed, and reformer Mikhail Gorbachev came to power. When Gorbachev visited Washington, DC, in 1987 to meet Ronald Reagan, David Harris, then director of the American Jewish Committee's Washington office (and later head of the whole AJC), organized a rally. Though the tide was already turning in favor of Soviet Jews who wanted to emigrate, a quarter of a million people turned up. They were, for the most part, not activists but American Jews who felt invested in this cause. At the rally, they called on Gorbachev to let their people go and sang "Hatikvah," Israel's national anthem[15] (unabashed support for Israel could also be seen in the same vein as this new rush of Jewish nationalism, felt and celebrated by American Jews who were newly unafraid of being cast as outsiders or accused of having dual loyalty).

Over 280,000 Soviet Jews came to the United States from the mid-1960s through the early 1990s.[16] Congress established a special refugee status for religious minorities from the Soviet Union,[17] which in turn meant that, from 1988 to 1993, in the Soviet Union's last years and the immediate aftermath, Jews from the Soviet Union constituted the largest refugee nationality to enter the United States. In 1992, nearly 15,000 refugees settled in Brooklyn, New York, alone—Brooklyn also being the most popular destination for Soviet Jews.[18] And once they arrived, American Jews, notably through HIAS—the Hebrew Immigrant Aid Society—worked to get them settled and situated.

According to the 2013 Pew study on American Jews, Jews born in the Soviet Union accounted for a full 5 percent of the American Jewish population that year.[19] Soviet Jewish refugees became Americans. But they did so with the background of their own experiences, which were in many ways distinct from those whose families had been in this country for decades.

As Roman, one young Jewish man from the former Soviet Union, told me, "You have this expectation because Jews have, for a long time, been kind of, I would say, more successful than a lot of immigrant groups in this country . . . there's this kind of expectation that you're going to be able to do well in school, get a bachelor's, get a master's degree, find a job in law or finance or medicine or something along those lines.

"I didn't take that path in life," he said. "Until very recently I was still trying to feel my way into being a stable adult." Roman spoke of a distance between himself and the mainstream narrative of those Ashkenazic Jews like me, whose families came over at the turn of the century.

"My experience has been more in line with an immigrant from another place in the early 1990s," he said.[20]

No group is a monolith, and no group's politics are monolithic, either. HIAS now helps resettle other types of refugees, stressing, "We used to take refugees because they were Jewish. Now we take them because we're Jewish." But there is a reluctance on the part of some Soviet-born Jews to see newer immigrants as anything but a security threat to their adopted country. Speaking to the *Atlantic* in 2016, Sam Kliger, then the director of the AJC's Russian-Jewish Community Affairs, conceded to journalist Olga Khazan (who was herself born a Soviet Jew who came to the United States when she was young) that that was not so surprising. "Every immigrant group wants to be unique," he said. "They come here, and they don't want others."

And as the Democratic Party's left flank became emboldened, with some candidates even identifying as Socialist, some Soviet-born Jews recoiled. In 2016, preliminary data from a survey conducted by the Russian-Jewish Community Affairs at the American Jewish Committee found that between 60 and 70 percent of Russian-speaking Jews were planning on voting Republican in the election; that is roughly the same percentage of American Jews overall who vote Democratic. Bernie Sanders may have been the Jewish candidate in the 2016 presidential primary, but, as Khazan wrote that year, "some in the community are recoiling from Bernie Sanders and his leftist ideals. One hundred years after the Bolshevik Revolution swept Communists into power, some Russians in America say they can't believe a serious candidate in the United States is calling himself a socialist."[21] (This is not dissimilar to the story of Cuban-born Jews, who similarly fled to the United States from Fidel Castro's Socialist government and so tend to vote more conservatively than American Jews at large. This is obviously not true of every Cuban-born Jew. The first Sephardic Jew to be in a US president's cabinet (if one does not count Judah P. Benjamin in the Confederacy) was Alejandro Mayorkas, whom President Joe Biden, a Democrat, selected to be his Department of Homeland Security secretary.[22] But, as Laura Limonic writes in *Kugel and Frijoles: Latino Jews in the United States*, Cuban Jews "have a history of voting for politically conservative candidates, largely stemming from an anti-Castro sentiment," whereas recent non-Cuban Latino Jewish immigrants are overwhelmingly Democratic.)[23]

Relatedly, Israel took on a place of privilege in the mind of many Soviet-born Jews. In the Soviet Union, theirs may have been a derided and resented nationality, but Israel, for some, could serve as a kind of fantasy homeland. Israel's victory in the Six-Day War instilled in some a sense of pride as Jews that the regime itself tried to deny them.[24] Roughly a million Russian-speaking Jews went to Israel after 1989[25]—many Russian-speaking Jews in the United States have relatives there; some even passed through themselves.

"We have a lot of relatives and friends and family in Israel because of the way that emigration from the former Soviet Union worked," Rachel, who asked that I refer to her by only her first name, told me. "The ties are personal." There are, of course, exceptions. Rachel herself said she has grown disillusioned with the country. "I don't know if I've changed, Israel's changed, or we've both changed," she said, but it feels like the country is "creeping toward annexation, and I can't support that."[26] Gennadyi Gurman, who moved to Israel when he was eight with his mother and then, as a teenager, to Brooklyn

with his father and brother, said that, over the past decade, he's had a
"mental divorce" between his Jewish identity and Zionism. "Because
you're supposed to have this loyalty to this place where theoretically
Jews are safe and have self-governance, you can't criticize any aspect,"
he said, drawing a comparison to those who demand uncritical love
for the United States.

Still, for some, supporting Israel wasn't just a part of being Jewish;
it was what being Jewish meant. "When I say negative things about
Israel, a lot of my family thinks I'm a Bad Jew," Gurman said.[27]

Once, Olya told me, her mother came to visit her and brought
an Israeli friend along with her. Watching them argue, she said, was
interesting. The friend hated Benjamin Netanyahu, the then prime
minister of Israel. Her mother, by contrast, supported Netanyahu
and his policies.

Olya herself disagrees that Jews around the world are bound by
some creed to support Israel unquestioningly. She told me that she
stopped talking politics with her parents "ages ago."

"Half of Israelis don't like Israeli policies," she said. "So the idea
that American Jews are more or less Jewish based on views of politics
in a country they don't live in is crazy."

Her own thought, she added, is that "Israeli politics is the business
of Israelis. I can have opinions," she said, "but the only country I get
to vote in is the United States."[28]

In some ways, the story of Persian, or Iranian, Jews is not dissimilar;
in others, it is wholly unique. (Both "Persian Jews" and "Iranian Jews"
refer to Jews whose families are from Iran. In this section, I refer to
individuals based on the term that they used to describe themselves.)

Iranian Jews can trace their lineage in Iran back hundreds, and indeed thousands, of years—all the way back to the start of the Persian
Empire in 539 BCE.[29]

But with the Islamic Revolution of the late 1970s, in which Mohammad Reza Pahlavi, the last shah of Iran, was ousted, thousands fled to America. Some came over before. Esther Amini is the author of *Concealed: Memoir of a Jewish-Iranian Daughter Caught Between the Chador and America*. She grew up in Queens, New York, in the 1960s, her family having already fled Iran. It wasn't as though there was no repression against Iran's Jews, no reason to flee to somewhere else, prior to the revolution.[30]

But it was the revolution that marked a turning point. Haideh Sahim, who previously taught Persian language and literature at Columbia University and NYU, told me, "I was born in a family both observant and not." They were "very strongly Jewish," but not particularly religious or observant. Sahim came to the United States for graduate studies. Then, "the revolution happened. Everything went completely changed."

She had a "very difficult time bringing [her] whole family over here," she said. (She was not alone. HIAS, which worked to resettle hundreds of Iranian Jews, noted a steady stream of Jews from Iran following the revolution.)[31]

Just as there is no one American Jewish experience, so was there no one Persian or Iranian Jewish experience in the United States. Many went to Los Angeles, California. Many others started their American life in New York City, where the first generation of arrivals started out, and helped others start out, in the jewelry business.

From New York City, some moved to Great Neck, a town on Long Island. Great Neck has a heavily Jewish population—but that population is, broadly speaking, divided between Ashkenazic and Persian Jews.

One young woman from Great Neck who asked to be identified by only her first name, Julia, said that Great Neck has "a very strong, flourishing Persian Jewish community."[32] I grew up in the town next to Great Neck. My doctor, my dentist, and my orthodontist were all there, a ten-minute drive away, and some of my favorite restaurants and shops were, too. And it was in Great Neck that, as a child and teenager, I heard otherwise liberal Ashkenazic Jews, who, I am sure, would never have spoken this way against any other ethnic group, rail against Persian Jews. They were tough to do business with, some complained. They *haggled*, others said, even over postage stamps and groceries. (That this is a stereotype applied to American Jews as a whole by their neighbors was evidently overlooked.) In a 2010 song called "Nassau (County) State of Mind," a parody of Jay-Z's "Empire State of Mind" and mostly loving send-up of Jewish suburban life, one young man raps, "I wanna move to Great Neck, become an oral surgeon. Can't move there, though, 'cause I don't speak Persian."[33]

To be clear, not every Ashkenazic Jew in Great Neck spoke this way about their Persian Jewish neighbors. But there was an element of division between Ashkenazic and Persian Jews, and not only because of Ashkenazic prejudice. They were living different experiences.

At home, Julia said, "everyone gets together all the time. If, God forbid, everyone in my family died, there would be a hundred families around me ready to take me in." She has come to love her close-knit community, though conceded, "It's good and bad. We're all up each other's asses all the time."

That's particularly true, she said, of families like hers—Mashhadi Jews, or Jews whose families come from the Iranian city of Mashhad. Julia explained that Jewish families in Mashhad were particularly oppressed, and so, here in the United States, they are particularly religious (the "darker side," she said, is that some are "extremely Islamophobic"). "I grew up in an ultrareligious community," she said, though her family in particular was less religiously observant.

As if to illustrate her point she said that, though she's getting married at twenty-two to another Mashhadi Jew, and having an Orthodox wedding, and has big Shabbat dinners with her family every week, she doesn't keep kosher and isn't *shomer Shabbos* (meaning that she will do things like use electricity, spend money, or write on Shabbat).

And so "my friends always make fun of me and call me a goy," she said. "I'm the goy friend."[34]

Not every Persian Jew moved to the United States and doubled down on religious observance. Indeed, Saba Soomekh, author of *From the Shahs to Los Angeles: Three Generations of Iranian Jewish Women Between Religion and Culture*, has said that religious pluralism is one of the marks of the American Persian Jewish experience. "In Iran, you only had one type of Judaism: traditional," she told NPR in 2014. "Iranian Jews found themselves in Los Angeles and now they can pick which Jewish movement they believe aligns best with their religious practice."[35]

But some did become decidedly more religious once they lived in the United States. Here, unlike in Iran, they could worship freely. This is very different from, say, Jews born in the Soviet Union, who are less likely than their American-born counterparts to be religious (though there are, of course, exceptions to both rules).

"The community around us—the Jewish community here—has become more religious," Sahim said. She herself married a Christian man. He's a professor of Iranian history, and lived in Iran for four years, and so understood some of her culture. Her parents liked and accepted him. After her father passed away, it was her husband who insisted on saying prayers on Shabbat (he now blesses the wine; her brother recites the *motzi*, or the blessing over the bread).

But once, she recalls, she was very sick, and so went to synagogue. The person who gave the sermon said that if your children marry outside the religion, you should say kaddish for them.

One member of the Mashhadi Jewish community in Great Neck said that, if they were to marry outside the religion, it would cause a scandal for their family. In recent years, whom they could marry got still more limited; under Rabbi Yosef Bitton, a Sephardic Jew originally from Argentina (his mother is from Syria; his father, from Morocco), who came to lead this Mashhadi community in the early

2000s, Mashhadi Jews adopted the Syrian Jews' Takana, the edict against intermarriage—which includes marrying those who have converted to Judaism. "In America we are all very exposed to inter-marriage and I believe that the Takana will help the Mashhadi com-munity to overcome the perils of assimilation in the near future," he said in a 2013 interview.[36]

But in some ways, Sahim said, becoming more religious led Per-sian Jews, particularly the young ones, away from their roots, and embracing Ashkenazic norms in America.

"I am very annoyed that our Jewish tradition is being ruled by Ashkenazic rules," she said. These rules have permeated into Iranian Jews' lives, even though they were never a part of it before. Some Ira-nian Jewish Orthodox women have started wearing wigs, which they never did. Iranian Jewish restaurants that are open on Passover aren't allowed to serve rice, which traditionally they do eat, even though it's Ashkenazim who aren't supposed to eat kitniyot (legumes, grains, and seeds). The American Board of Rabbis errs on the side of Ash-kenazim. Parents who send their children to yeshivas become distant from them, Sahim said. The children are learning a different way of being than their parents.

"Why can't you recognize us?" she asked me. "We are older. At least the Iranian [Jewish] community is older—it's the oldest commu-nity outside the land of Israel.

"Respect us," she said. "We respect you."[37]

"After Joe Biden won the presidency, my liberal friends—mostly Ashkenazic Jews with deep roots in America—were aghast that over 70 million Americans voted for Trump," wrote Mijal Bitton—incidentally, Rabbi Yosef Bitton's daughter—in mid-November 2020. "My Syrian, Persian, Bukharian and Hispanic friends and family members—Jews with immigrant identities—were shocked, too. But most mourned the president's defeat."[38]

And this, too, was a difference between Great Neck's Ashkenazic and Persian Jewish communities. The former was overwhelmingly liberal. The latter went for Trump.

Not every Iranian Jew, of course, is a Republican. Anna Kaplan, a state senator from New York and Jewish refugee from Iran, is a proud Democrat.

When she was five or six years old in Iran, she went into the produce store—there weren't general grocery stores, she said, and customers would instead go to one store for bread, one for produce, and so on—and "the shop owner turned and said, 'Don't touch anything, tell me what you want.'" It came, she said, from the belief that Jews aren't clean.

In Iran, she said, "we tried not to attract a lot of attention to ourselves, not rock the boat, just live our lives, put our head down, and move on with it.

"I was really just grateful at age thirteen having an opportunity to come to this country," she said. "I was so excited to come here."

And though she comes from a "pretty conservative" family, her own American story is wrapped up in liberal values.

She always talks, she said, about how much this country has given her. In how many countries could you come here as a teenage refugee and become elected to the New York State senate? But she was able to do that, she told me, because she had opportunities. And it is "paramount," she said, "to make sure those opportunities do stay available for others.

"My religion teaches me to treat others the way I'd like to be treated by them. To see my neighbor as equal to myself."

Kaplan first ran for office "because I wanted to make sure Iranian Jews here understand the importance of registering, voting, making sure they understand they need to be part of the conversation. If they don't want to be . . . the decisions that will be made will be made without their input."[39]

But that doesn't mean that she agrees with their contributions to the conversation, or they with hers.

"Among Middle Eastern Sephardic Jews living in America, many share family histories of having escaped Arab nationalism and anti-Semitism in the Middle East," Mijal Bitton wrote in her JTA piece. "Their recent experiences of Jewish displacement have led many to identify with a realpolitik approach in which a 'strongman' politician can best compete in the international arena to protect both Israel and American interests. Moreover, many are socially conservative and identify with Trump's economic policies."[40]

Kaplan, when I spoke with her, echoed roughly the same sentiments.

"A lot of people in my district, the Jewish Iranians, I wouldn't even call them Republican. I would call them Trumpist. They believed in President Trump." If you look at an electoral map of the north side of Great Neck, which has a high concentration of Sephardic and Mizrahi Jews, she told me, it's "dark red." (I did look at such a map, and it does show that between 70 and 80 percent of the heavily Persian Jewish area went for Trump in 2020.)

"I think the majority will probably tell you they believe that way because they felt President Trump was a friend to Israel." They are fearful that Democrats will be swallowed by the extreme left and not support Israel enough, or not support it at all. (She added that she also thought it was about their "pocketbook," another way of saying the Long Island Iranian Jewish community is more economically conservative.)

Kaplan herself, though not a Republican, is extremely vocal about her belief in Israel. In her conversation with me, she spoke adamantly against the idea that the term "apartheid" could be used to describe Israeli policy, as some human rights groups and progressive politicians say. When people are only "picking on Israel," she said, "and not talking about genocide in China or Syria, it is antisemitic."[41]

The Democratic debate on how to criticize Israel will be examined further, but for now it is worth noting that newer Jewish immigrant groups—like Soviet Jews and Persian Jews—can tend to be more politically conservative, and certainly less willing to allow criticism of Israel.

It is impossible to pinpoint one reason. Writing in the *Forward* in 2020, Philip Mehdipour, an Iranian American Jew, outlined why he believed his own community supported Trump. Of the revolution that led them to flee Iran, Mehdipour wrote, "Iranian Jews blamed two culprits for the tragic collapse of their home: The first was the Ayatollah Ruhollah Khomeini, the head cleric who imposed a reign of religious terror that continues to this day. But Iranian Jews also blamed President Jimmy Carter. Carter's human rights policies gave life to a dormant democratic movement that was dominated by Khomeini. . . . His rants against women's rights and land reform had gone unnoticed, and he was ironically able to assume the mantle of a Democrat, a progressive cleric reformer fighting against despotism and colonialism." Certainly, it could be that, as some Soviet Jewish refugees and their children credit Reagan and associate Democrats with the repressions they experienced in the name of Communism, so do some Iranian Jews still vote Republican because they don't trust the Democrats after Carter.

But it is also worth noting that both of these groups are not as far removed from their own families' traumas, their own escapes, as families like mine. And Israel provided a refuge from that trauma. "Israel currently has the largest population of Iranian Jews in the world. Israel granted Iranian Jews citizenship during their time of crisis in 1979. So it's no surprise that they would be defensive of the country," Mehdipour wrote. (He added that Iranian Jews resented the Iran nuclear deal that President Barack Obama negotiated with Iran, in which sanctions on Iran were lifted in exchange for Iranian pledges to curtail its nuclear enrichment program, and which came into effect in 2016; Trump withdrew from it the next year.)[42]

For most American Jewish families, that moment of escape came earlier on. "Now [you're] a couple of generations past that. If all your parents and grandparents are born and raised in America, that's all you know," said Maayan Malter. Malter grew up in the United States but was born in Israel; her mother is half-Israeli. Many American Jews, she said, "don't have direct connection to the old world, this fear." In her own case, though, her mother's side of the family escaped from Iraq. That directly informed the stories she grew up with and her way of looking at the world. "Even if we feel safe, we always have to be a tiny bit cautious. . . . you see that in Persian communities," she said, describing many Persian Jews as "super right-wing, protective of Israel. . . . That's their experience."[43] Indeed, in my conversation with Julia, she told me that, as observant as her family was, her father always taught her that that wasn't what really mattered; what really mattered was Israel.[44]

And here is a potential paradox for the contemporary liberal American Jew. The American Jewish community is majority white (or, if one rejects the idea that Jews can be white, "white passing," and going through life with the privileges of a white American) and majority Ashkenazic. The term "Ashkenormativity" has been described as the assumption that all Jews are Ashkenazic, refusing to imagine or consider Jews of other backgrounds—indeed, in the American context, it can be used to sub in a certain cultural shorthand for all of Judaism or Jewishness (though, as Jordan Kutzik, a Yiddish children's book publisher, put it to me, what we think of as "Ashkenormative," like referencing bagels and lox and *Seinfeld*, isn't actually particularly Ashkenazic; "actually learning Yiddish isn't Ashkenormative at all," he said).[45] Ashkenormativity can also leave out the experience of Ashkenazic Jews who came over later.

Liberal, progressive Jews rail against this kind of exclusive, limited understanding of Jewishness. "Ashkenormativity Is a Threat to All Jewish Communities," read the headline of an article written by Isaac Ofori-Solomon, an Ethiopian Jew, and published by Hey Alma, a feminist Jewish pop culture site, in 2020. And it is. It flattens the Jewish experience and leaves out American Jews. There are, today, American Jews pushing back on Ashkenormativity in meaningful (and less meaningful) ways.

But part of accepting different types of Jewish people means accepting different types of Jewish politics as Jewish, even while disagreeing with them, and even as they are at odds with what we think of as the politics of being more progressive and inclusive. As Mijal Bitton wrote in her JTA piece, "if we want to truly understand those who hold views different than ours, we must take the time to get to know diverse communities rather than ascribing our own political beliefs and assumptions onto others."

Before I hung up with Sahim, I asked if there was anything else she wanted to share—something that hadn't come up in conversation but that she thought was important.

She said she would like the Iranian Jewish community "to try and be more—I don't want to say assimilated, but accultured—to be a little more open to the American way of life.

"Not so much that they will lose their own way of life," she added. "Like the Italians. They're very Italian, they still eat their own food, but they're also American.

"I would like to see more of that. And that is possible."[46]

On the other hand, some look at the distinctiveness that Persian Jews have been able to maintain admiringly. Vladislav Davidzon came to the United States as a child from the former Soviet Union. He is the son of Gregory Davidzon, onetime media mogul and political kingmaker of Russian-speaking Brooklyn. Vladislav made clear that he did not think the United States was a bad country. "I'm grateful to this country," he said. "I've had and squandered opportunities that other people can only dream of."

But Vladislav told me he does not live mostly in the United States, instead splitting his time between Paris and Kyiv, working as a cultural correspondent for the Tablet, an online Jewish magazine (our interview was before Russia went to all-out war in Ukraine in 2022; after it did so, Davdizon went to Ukraine to report). Today, he said, he lives in the rarefied, stratified postmodern world of Russian and Jewish wealth.

"I don't like American Jews," he said. "I don't really want to be one." American Jews, he said, gave up a lot of what he finds interesting about Jewish people in order to become American. "I find a lot of them to be weak, conformist." He is on a roll now. American Jews, he said, gave up their cultural characteristics. They did not pass their cultural capital on to their children and grandchildren, though, he allowed, the Persian Jewish Diaspora did.

"The people that I like all have interesting lives," he said. "I do feel there's something really deeply boring about ordinary American Jewish striving."

Assimilation went "horribly wrong" in his case, he told me. And I did not doubt that he is happier and has found more appealing cultures elsewhere. It did amuse me, though, that, despite himself, he was engaging in a very American Jewish practice: harshly critiquing American Jews of the present and looking to other times and places in search of other, more interesting Jews.[47]

American Jewish history is a history, or a set of histories, of immigration and the subsequent oscillation between accepting and resisting acculturation. Of learning how to be American and something else at the same time, and of figuring out what that means.

And very often, though not always, that immigration was bound up in escape, or a seeking of refuge, or a flight from persecution.

Despite this, there have been, in very recent US history, American Jews who have distinguished themselves by actively working to restrict immigration to the United States—and, not unrelatedly, American Jews who have distinguished themselves by working to police the borders of American Jewishness.

Chapter 8

"This Land Is Our Land" Jews

AMERICAN JEWISH HISTORY IS entangled with American immigration history. That does not mean, however, that every American Jew welcomes or celebrates or fights for immigration.

Some have. Some, like Mannie Celler, the congressman who led the charge for immigration reform, advocated for a more expansive and inclusive approach to immigration (at least in theory). Others, like HIAS, moved from advocating for Jewish refugees and immigrants to advocating for refugees and immigrants more broadly.

Theirs is one set of stories. There are others. There are the Jews who lobbied Congress to continue to be thought of as white, not "Asiatic," and who declined to press for immigration for Asian people so long as they, Jews, could still come to and assimilate in America. There were Ashkenazic American Jews who pushed off their Sephardic coreligionists. There were the American Jews who didn't want to advocate too forcefully for European Jews in the years after Hitler came to power but before the Holocaust. And there were Jews who actively worked to enact Donald Trump's anti-immigrant agenda.

Donald Trump's presidency marked a unique moment in American Jewish history. On the one hand, he openly disavowed many of the liberal values that most American Jews have historically supported at the polls. On the other hand, he tightly embraced then Israeli prime minister Benjamin Netanyahu, lambasted Democratic politicians who criticized Israel, moved the US embassy to Jerusalem,[1] and oversaw the Abraham Accords, under which certain Arab countries diplomatically recognized Israel.[2] The overlap between Trump and Netanyahu was particularly striking since Trump's predecessor, Barack Obama, despite signing off on a decade-long military aid package worth $38 billion,[3] clashed with Netanyahu, crafting the Iran nuclear deal against the Israeli prime minister's objections[4] and having his ambassador to the United Nations, Samantha Power, abstain on instead of veto a December 2016 United Nations resolution condemning Israeli settlements.[5] If being "pro-Israel" means offering Israel uncritical support, then Trump was an unabashedly "pro-Israel" president. Israelis thought so, too: unlike American Jews, they overwhelmingly supported him.

This chapter is about two types of people. It is about the Jews who supported Donald Trump in trying to keep out foreigners and paint as un-American those who stood in the way of his agenda. And it is about the Jews who tried to paint as not Jewish those who took issue with the former president and his Israel policies.

Donald Trump announced his presidential campaign on June 16, 2015, with a speech and rally at Trump Tower in New York City. Immigration featured heavily in his announcement.

"When Mexico sends its people, they're not sending their best," he said. "They're not sending you," he said, drawing a contrast between his audience and the people he was promising to keep from them. "They're sending people that have lots of problems, and they're bringing those problems with us. They're bringing drugs. They're bringing crime. They're rapists. And some, I assume, are good people.

"I would build a great wall, and nobody builds walls better than me, believe me, and I'll build them very inexpensively, I will build a great, great wall on our southern border. And I will have Mexico pay for that wall," he vowed.

His criticism was not reserved for President Barack Obama's administration. Of one of his Republican primary opponents, Jeb Bush, Trump said, "He's weak on immigration."

"Sadly, the American dream is dead," the speech concluded. "But if I get elected president, I will bring it back bigger and better and stronger than ever before, and we will make America great again."[6]

The project of making America great again was explicitly and directly linked to preventing immigration, but not all immigration: in 2018, as president, Trump said that he would have liked more immigrants from countries like Norway.[7] But immigration from Latin America, South Asia, Africa, and the Middle East was presented by Trump as threats not only to the security but to the very state of the nation.

This language—this focus on the wrong kind of immigrants—was not dissimilar from that which saw Jewish immigration restricted in the early twentieth century. Nevertheless, Trump found prominent American Jewish partners in his project. His eldest daughter, Ivanka Trump, who introduced her father the day that he announced his candidacy and went on to work in his White House, is Jewish; she converted to Judaism shortly before marrying Jared Kushner, who, as the president's son-in-law, worked in the Trump White House and was tasked with establishing peace in the Middle East (he was also a family friend of Netanyahu).[8] In the 2020 vice presidential debate, Vice President Mike Pence said to Kamala Harris of the man at the top of his ticket, "Your concern that he doesn't condemn neo-Nazis—President Trump has Jewish grandchildren. His daughter and son-in-law are Jewish."[9] (As was true in the case of Henry Ford and Charles Lindbergh, being related to or having a close relationship with a Jewish person does not, in fact, mean that someone is not an antisemite or white supremacist.)

But perhaps the most notable Jewish person to serve in the Trump administration was Stephen Miller, who, as Trump's senior adviser for policy, worked aggressively to craft and champion Trump's draconian immigration policies.

Stephen Miller had a maternal ancestor by the name of Wolf-Leib Glosser, who arrived in the United States shortly before the Kishinev pogrom. With help from HIAS, he and his brother, Nathan, were eventually able to bring their family over from the Russian Empire, as Miller's uncle, David Glosser, recounted to journalist Adam Serwer in the latter's *The Cruelty Is the Point*.[10]

As David Glosser has said, under immigration policies championed by Miller, his ancestors—poor, unskilled, non-English-speaking—would not have been allowed into the United States. In 2018, Glosser, who volunteers with HIAS, wrote in *Politico* in an article titled "Stephen Miller Is an Immigration Hypocrite. I Know Because I'm His Uncle," "I shudder at the thought of what would have become of the Glossers had the same policies Stephen so coolly espouses—the travel ban, the radical decrease in refugees, the separation of children from their parents, and even talk of limiting citizenship for legal immigrants—been in effect when Wolf-Leib made his desperate bid for freedom."[11] This did not sway Miller.

But perhaps the most telling moment related to Miller came not from anything his uncle said but from his own mouth. In 2017, while defending a new "skills-based" immigration program, Stephen Miller was asked by CNN reporter Jim Acosta, "The Statue of Liberty says, 'Give me your tired, your poor, your huddled masses, yearning to breathe free.' It doesn't say anything about speaking English or being a computer programmer. Aren't you trying to change what it means to be an immigrant coming into this country if you're telling them that you have to speak English?" To this, Miller replied that, since naturalization requires people to speak English, the idea that English isn't involved in immigration processes is ahistorical. He went on, "Secondly, I don't want to get off into a whole thing about history here, but the Statue of Liberty is a symbol of American liberty lighting the world. The poem that you're referring to was added later (and) is not actually part of the original Statue of Liberty."[12]

And Miller was right. The poem was added later. It had never intended to be a symbol of immigration. But Emma Lazarus saw the call for the poetry contest and thought of her own Ashkenazic and Sephardic ancestors. It was an American Jew who decided that the statue was a symbol of immigration, just as it was an American Jew who decided that, actually, he didn't think it should mean anything at all.

And who is right? Which is the real Jewish response?

Lazarus's Jewishness is certainly more appealing to me personally. It is comforting to tell myself the story of the American Jewish woman who created out of a statue a symbol of American openness to immigrants. I can look at Lazarus's poem and at my own synagogue's work to help refugees and tell myself that it is all part of the same Jewish narrative. And it is a true story, a real episode from Jewish history.

But so is Miller's.

Ben Lorber, a research analyst at Political Research Associates who works on antisemitism and white nationalism, told me, "We liberal Jews, broadly defined, might be in a bubble.

"Jews are human beings," he said. "Like any other group of people, there are going to be members of our community who fall on all sides of the political spectrum." He comes into contact, in his work, with Jews who are supporters of full-fledged white supremacist movements. ("Never mind Stephen Miller," he said. "These are outright white nationalists.")

"I think about how these people are supporting policies that would have led to their great-grandparents being trapped in Europe as Hitler was advancing," he said. "There's a part of me that wants to scream, '*Shande*, how could you!'"

But when I asked him what came to mind when he heard the term "Bad Jew," Lorber, who previously worked as a campus organizer at Jewish Voice for Peace, a group dedicated to liberation and justice for Palestine, and which is itself often maligned as being full of self-hating Jews, told me that he recoiled at the term, and that he strives for a pluralism and unity where all Jewish identities are considered valid, "extending the same . . . to Jews on the right whose policies I radically disagree with.

"There are enough voices in the world who are labeling Jews as bad," he told me. "We shouldn't do it to each other."[13]

I thought of what Lorber said while on a tour of Jerusalem's Old City. My guide, Dr. Ian Stern, an archaeologist who made aliyah some four decades ago, showed me a piece of what would have been a mansion in the Upper City in the year 70. It had been burned.

The Second Temple was destroyed by the Romans on the ninth of Av (Jews now mark the day with Tisha B'Av). Roughly a month later, on the eighth of Elul, Josephus, a historian writing at the time, recorded, "Of the rebels, some already despairing of the city, retired from the ramparts to the citadel, others slunk down into the tunnels. Pouring into the alleys, sword in hand, they massacred indiscriminately all whom they met, and burnt the houses with all who had taken refuge within." "They" in this sentence is often taken to be the Romans—on the Israeli Foreign Ministry website, for example, it is specified that it is the Romans to whom Josephus is referring, who burned the Upper City houses.[14] But my guide told me that, actually, we can't know who "they" is, and whether it was the Romans or the

Jewish rebels (or "zealots," as they are sometimes called) who burned the mansions, disgusted at the upper-class, holier-than-thou neighbors living comfortably in this time of peril. My guide invited me to think of it as relevant today: the idea that, surrounded by Romans, Jews could turn on one another.

I thought about that for a while. I wondered, *If it was the case then that Jews attacked Jews, what is the parallel for Jews in today's world? Who is throwing the flame, and what is getting burned down?*

The cost of Trump's immigration policies was great. The American Civil Liberties Union identified more than fifty-five hundred children as having been separated from their families at the border.[15] Over forty immigrants died in US Immigration and Customs Enforcement custody during Trump's time in office. His predecessor, Barack Obama, had a decidedly mixed record on immigration, at once championing those who came as undocumented immigrants as children and also earning the moniker "deporter in chief." Still, before December 2018, it had been more than a decade since a child died in US Border Patrol custody.[16]

And it was not only immigrants who felt the consequences of a president who ran and governed with racist rhetoric. Though correlation doesn't equal causation, one could surmise that it is not accidental or coincidental that hate crimes surged by more than 20 percent, hitting their highest point in nearly a decade, during the Trump presidency.

American Jews were not the primary targets of Trump's ire, but neither were they spared from its aftershocks. In 2017, I watched, at once dumbfounded and unsurprised, as men holding tiki torches marched on Charlottesville, Virginia, and chanted, "Jews will not replace us." I watched the next day as the president said that there were "very fine people on both sides." Gary Cohn, his Jewish chief economic adviser, did not resign. He would later quit, but not until 2018, after Trump announced steel and aluminum tariffs. That day in 2017, he did not rush the microphone and yell that his boss was wrong and that he was wrong for being there, for serving this person.[17]

My father, when I called him that day, said, "And another thing. These people with the torches: You think I want to replace you? I'm good, thanks." And I laughed back.

But the truth was that I was tired of laughing. I was tired of Jewish jokes and jokes made by Jews to deflect against Jewish jokes. I understood, and understand, that it helps some people. That it allows them to feel like they are taking the power out of conspiracy theories and hatred. I am not saying that they shouldn't do that. There is a long Jewish tradition of doing so.

But I didn't want my dad to feel like he had to joke that he didn't want to take the place of the sad white supremacists wielding tiki torches. I wanted the white supremacists to not have done the chant and the march in the first place. And failing that, I wanted to say that I thought they were repulsive and small and sad and could get fucked. They weren't joking. Why should I?

Many American Jews felt that day in 2017 to be a dark one. Darker was still to come.

One of the tropes that Trump pushed was that immigrants were flooding the country. They were coming in in numbers too great to control and were going to undermine the country as it was known and loved by his followers. If you don't have borders, you don't have a country, he said. This is how he positioned immigration: as an existential threat to the United States.

As the 2018 midterms approached, Trump, trying to whip up anti-immigrant fervor, focused on migrant caravans at the country's southern border.

On October 27, 2018, a man armed with an AR-15-style assault rifle and at least three handguns and antisemitic slurs entered the Tree of Life synagogue in the Pittsburgh neighborhood of Squirrel Hill. Before surrendering to the police, he killed eleven people. It was the deadliest assault on Jewish people in the United States in the country's history. Afterward, it was discovered that the shooter, before murdering eleven Jews, had posted on social media about HIAS, which, he said, "likes to bring invaders in that kill our people."[18] The Jewish Family and Community Services of Pittsburgh was one of HIAS's resettlement partners.[19]

A few days later, with some prompting from a reporter, Trump said he "wouldn't be surprised" if it were Hungarian-born philanthropist George Soros who was funding the caravan.[20] Though Soros's Open Society Foundations has given money to back immigration support, it has never funneled money to migrant caravans.[21]

Jennifer, who asked to be referred to by only her first name, grew up in Pittsburgh. Though she now goes to a different synagogue, she had her bat mitzvah at the Tree of Life synagogue and lives around the corner. "Many of the people who died on that day were people I knew," she said, including the mother of one of her best friends.

What happened was shocking and unsettling, she said, but so, too, was she upset by what came after. "People just became so politicized," she told me. There were anti-Trump protests, blocking the street, preventing her children from going to Hebrew school, robbing them, she said, of a sense of normalcy, routine, and community at a time when those things were especially important to them.

"We haven't even buried the dead yet," she remembers thinking. "Can we just not make this about him?

"But people are angry," she said, "I get that."[22]

I saw a connection between Trump's rhetoric and the violence carried out at Tree of Life. I was angry, too, in the days that followed when he visited the synagogue. *You did this to us*, I thought. And if we were making it about politics, surely politics had first taken an interest in us.

But I didn't live in the community. I wasn't there, grieving people I knew, hoping to see people come together, only to have my heart broken at realizing they wouldn't or couldn't.

There were other divisions that existed not within Squirrel Hill, but around it. Hannah Lebovits grew up in Pittsburgh to a mother who grew up in, as Lebovits put it, what would have been called a modern Orthodox home and a father whose parents were Holocaust survivors from Satu Mare, then in Hungary. Hers, she said, is part of a growing demographic of American Jews that has moved further to the right.

She went to Yeshiva Achei Tmimim, the local Chabad cheder, for elementary school and went on to get an undergraduate and graduate degree; she has six siblings but was the only one, at the time of the interview, who went to graduate school or has what, in her words, would be considered a career. Her spouse is Lubavitch (Lubavitch Hasidism being a branch of Hasidism that traces back to a townlet known in Yiddish as Lubavitch, which was a center of Chabad Hasidism). Together, they decided they wanted to work in more secular settings. Lebovits is now an assistant professor at the University of Texas, Arlington, working "at a top university in a country with a family full of people who have no idea what I do, no idea what I'm saying, ever."

Before that, however, she once studied at BJJ Seminary, a Haredi girls' seminary in Jerusalem. She had a Google email group with her cohort from her seminary. After the Tree of Life shooting, she said, she noticed nobody was talking about it.

"I sent an email to the group saying, 'I'm from Pittsburgh, you all know, how come so few of you have reached out,'" she said. "I got a few responses essentially saying, 'we just assumed that this wasn't a religious synagogue, so probably people there weren't really Jewish.'"[23]

In early 2022, a British Pakistani man came to Texas and took several people—including the rabbi—hostage at Congregation Beth Israel synagogue in Colleyville. The synagogue is Reform. When it was over, and all the hostages had gotten out safely, I received an email from Lebovits. "I was struck once again by the lack of any sort of communication from many of my religious friends, in light of yesterday's tragedy which hit so close to my Dallas home," she wrote.

This mentality is not unique to Lebovits's religious friends. In Israel, the traditionally Orthodox media refused to refer to Tree of Life as a synagogue because Tree of Life is Conservative. Asked to comment, Israel's Ashkenazic chief rabbi, David Lau, appeared to equivocate, deeming it instead a "place of clear Jewish character."

In fairness to Lau, though, he also said, "I have a deep ideological disagreement with them about Judaism, about its past and the consequences for the future for the Jewish people for generations. So what? Because of that they are not Jews?"[24]

Raising the question rhetorically, though, doesn't change the fact that there are some who would answer "yes."

There is another current to all of this, which is that there are some who insist that Trump was, in fact, good for Jews because of his policies toward the state of Israel. Some of the mainstream, establishment Jewish institutions, like the American Israel Public Affairs Committee (AIPAC) and the American Jewish Committee, were accused of pulling their punches on Trump, choosing to focus instead on leftists who spoke out against Israel.[25]

Daniel Elbaum, AJC spokesperson during the Trump years, told me that the AJC "has always tried to work on the particular and the universal," advocating for both Jews and democratic values. He added, "No matter who the occupant of the White House is, there is always an inherent tension between maintaining a level of access to those making the decisions and speaking out publicly and strongly on every single occasion." Some have decided that they do not care about having access to the White House, he said, but that is not where the AJC was.[26]

The Anti-Defamation League was similarly criticized, with people from within the organization telling reporter Alex Kane that the organization had positioned its long-standing commitment to civil rights as second to support for Israel, even if that meant pulling punches on Trump.[27] (This is to say nothing of groups like the Zionist Organization of America, the leader of which, Mort Klein, tweeted that Black Lives Matter "is a Jew hating, White hating, Israel hating, conservative Black hating, violence promoting, dangerous Soros funded extremist group of haters" in 2020,[28] despite, among many other issues, the ADL's warnings that alleging that Soros is all-controlling is antisemitic.)

There are also some who argue that those American Jews who do not feel Trump is best for them are either deluding themselves or not taking seriously the concerns of more observant—and specifically traditionally Orthodox—Jews, who are made vulnerable to antisemitic attacks by virtue of being more "visibly" Jewish and who did indeed overwhelmingly support Trump[29] (studies have shown that modern Orthodox Jews were more evenly split between political parties).[30] While traditionally Orthodox Jews may not have necessarily agreed with Trump's approach to, say, immigration—some went above and beyond in rescuing Afghans as the United States withdrew and the country fell to the Taliban in the summer of 2021[31]—they supported enough of his policies, and enough of what he said and did, to vote for him in large numbers.

"Orthodox Jews see @realDonaldTrump as super friendly to Jews and great for Israel and a real fighter against antisemitism. Non-Ortho (secular, unaffiliated, Reform, Conservative, Reconstructionist, whatev) see him as an enemy. Any way to tell whose gut instinct may be better?" Jeff Ballabon, an American media executive and consultant and Orthodox Jew sometimes characterized as the architect of "Jexodus" (the idea that Jews will leave the Democratic Party en masse), tweeted in December 2019.[32] He went on: "Well, one way might be to compare who is most directly and most often actually impacted by antisemitism. Orthodox Jews—by a wide margin. Another way might be to compare who is more educated about Jewish history and religion. Again, Orthodox Jews—by a wide margin. Yet another might be to compare who demonstrates greater affinity and concern for the Jews of Israel. Yet again, Orthodox Jews—by a wide margin." I do not doubt or dispute that Orthodox Jews experience antisemitism that I never will. In recent years, traditionally Orthodox Jews have been attacked in kosher grocery stores and on the streets of New York City, and have been targeted simply for existing visibly and publicly. The implication in these tweets, however, was that Orthodox Jews had more authority as Jews, and so their gut instinct regarding whether Trump was "friendly" to Jews was "better." (The idea that Reform Jews are less Jewish is also an issue in US-Israeli Jewish relations. In Israel, where the default is Orthodox Judaism, the High Court of Justice only ruled to recognize Reform and Conservative conversions in Israel for the purposes of citizenship in 2021; the vast majority of applicants for non-Orthodox converts under the Law of Return that year were not approved.[33] Barak Dunayer grew up secular in Israel; in New York City, to hold on to his Jewish identity, he became a practicing Orthodox Jew. He is also my parents' real estate agent. Of Conservative and Reform Jews, he said, "Many of those Jews are great supporters of Israel, and you don't want to drive those people away. . . . And I'm saying that as an Orthodox Jew. I just think it's wrong to push away the others.")[34]

In some cases, the people who insisted that the true Jewish thing to do was to vote for Trump were themselves Jewish. When Trump said that Jewish Democrats were "disloyal" to Israel, and so was accused of playing with the old trope of dual loyalty, the Republican Jewish Coalition attempted to do damage control by saying that Trump was implying that Jewish Democrats were disloyal "to themselves"[35] (Trump's lawyer, Rudy Giuliani, helpfully clarified that the then president did indeed mean disloyal to Israel).[36] In other instances, the comments come from people who are not Jewish, as in the case of Trump, who, after he lost reelection, told the traditionally Orthodox publication *Ami* that he did as poorly as he did with American Jewish voters because American Jews "don't love Israel enough. Does that make sense to you?"[37]

The existence of individuals confused as to why American Jews vote for Democrats is not a new phenomenon. The *Commentary* crew was perplexed by the same. The quandary is perhaps not as confusing as all that. According to Pew's 2020 survey of American Jews, 83 percent of those who regularly attend services say they do so to feel closer to their history.[38] American Jews are also, still, despite the assimilation and acculturation, a minority. That a minority population keenly aware of its own history of suffering would vote to try to continue pluralism has never seemed to me to be especially surprising. Even my grandfather, who loved Israel and considered it core to his Jewish identity, did not necessarily cast his votes based on it, and indeed was elected to office in Boston as a Democrat. But it is, of course, the prerogative of politicians and their allies to wonder why people do not vote for them.

What is new, or has at least seemed to have intensified, is that those who believe American Jews should defect from the Democratic Party or show insufficient support for Israel are not content to wonder why they do not. Rather, they now insist that those who disagree with them are Bad Jews, or even not really Jews. And here we have another kind of Jewish xenophobia: finding foreigners—Jews whom they deem unworthy of belonging among Jews—in their own midst.

• • •

Perhaps the most famous example of this is the case of George Soros. Soros is the target of much criticism, some of which is wholly
fair: he is arguably history's most famous currency speculator, and
all his philanthropy has clearly not corrected the imbalance between
his wealth and power and that of his grant recipients. But some of
the criticism is less actual critique and more antisemitic conspiracy
theory. In 2018, Hungarian prime minister Viktor Orbán said, ahead
of Hungarian parliamentary elections, "We are fighting an enemy
that is different from us. Not open but hiding; not straightforward
but crafty; not honest but base; not national but international; does
not believe in working but speculates with money; does not have its
own homeland but feels it owns the whole world." His "Stop Soros"
campaign alleges that Soros is responsible for flooding Europe with
migrants and refugees. The perpetually foreign Jew who wants to
undermine the true nation is a classic antisemitic trope. (It was, of
course, later that same year that Trump suggested that Soros could
be behind the migrant caravans at the southern border.)[39]

But when told that these were antisemitic tropes, rather than
apologize, Hungarian officials and Trump allies alike said that they
couldn't be, because Soros wasn't really Jewish. Hungarian officials
pointed to his relationship with Israel as proof to question Soros's
Jewish qualifications (Soros, a funder of Human Rights Watch,
among other groups that support Palestinian rights, raised the ire of
Benjamin Netanyahu's government and Netanyahu's son Yair, who
accused Soros of destroying Israel).[40] Giuliani, who accused Soros
of working against Trump in Ukraine, said, in response to charges
of antisemitism, that he, Giuliani, who is not Jewish, is "more of a
Jew" than Soros, because "he doesn't go to church, he doesn't go to
religion—synagogue. He doesn't belong to a synagogue, he doesn't
support Israel, he's an enemy of Israel."[41] To note that, per Pew, only

35 percent of American Jews are in a household in which at least one person is a member of a synagogue is to miss the point,[42] which is that there are those who have taken it upon themselves to dictate that those who disagree with them are not really Jewish, and so antisemitic attacks against them cannot really be considered antisemitic, and their concerns as Jewish people can be dismissed because they are not really Jewish at all.

It is not only people who are not Jewish who take it upon themselves to dictate who is and is not really Jewish based on political positions or level of observance. In the spring of 2021, Alan Dershowitz, who wrote in the 1990s that Jews who intermarried did not have Jewishness at the center of their identities,[43] proclaimed that Senator Bernie Sanders, who called for a cease-fire between Israel and Hamas in Gaza and said that both Israel and Palestinians had a "right to live in peace and security," was a "self-hating Jew."[44] Bernie Sanders had spoken at length about how much being Jewish meant to him and how it influenced his outlook on the world while campaigning for president. This did not matter. He was not being Jewish as Alan Dershowitz believed he should be, or speaking the way in which Alan Dershowitz thought a Jew should speak, and so he was a self-hating Jew.

A self-hating Jew or a Bad Jew is still a Jew. This condemnation is still not far enough for some.

Eli Valley is an artist. His anthology, *Diaspora Boy: Comics on Crisis in America and Israel*, includes ten years of cartoons. One of Valley's most famous comics shows Sheldon Adelson shaking hands with right-wing symbol Pepe the frog. "So it's agreed," Pepe is shown as saying. "We exterminate the Jews here, and you can exterminate the Palestinians there." Adelson is depicted as replying, "Sign me up."

Valley's work is a send-up of mainstream narratives of American Jewish life and of Zionism. The book, he told me, is "a reclamation of non-Zionist and Diaspora strains of Judaism." Multiple people, on learning that I was writing this book, told me that I had to speak to Valley. His work meant so much to them, they told me. It had helped them figure out their own relationship to Jewishness.

In 2019, Meghan McCain, then cohost of noted daytime talk show *The View*, lambasted Representative Ilhan Omar for comments on the Israel lobby. In response, Valley drew a comic in which McCain is shown crying while wearing both a cross and a yellow star reading "Jude." "The things she said about the Holy Land," Valley's McCain said. "She wants to exterminate us Jews!" The comic was about Meghan McCain claiming the trauma of, and assuming the right to speak for, Jewish people. McCain called the comic "one of the most antisemitic things I have ever seen."

It wasn't just her. That year, Valley told me, "so many representative figures in the Jewish community were embracing Meghan McCain as a token Jew while delegitimizing and erasing my Jewishness."

I asked what it was like to have people—Jewish people—criticize him in such a way that aims to negate his very Jewishness. John Podhoretz, son of Norman Podhoretz, called him a capo, likening the Jewish artist to Jewish prisoners who helped run Nazi camps. "At this point, it's not new to me. I was getting these comments on *Forward* comics in 2009. Now we're talking a dozen years later.

"It's not that it doesn't affect me," he said. But he knows that there are also people like the ones I spoke to, who found themselves bolstered and supported and challenged by his work. "The fact that they exist is not just gratifying but a buttress when I get these hate mails or whatnot," Valley said. "I know I'm not alone."[45]

But he's not alone in more ways than one. Valley is a high-profile example, but there are, in fact, many Jews from whom other people, including other Jews, try to strip Jewish identity for having a certain set of politics or outlook on the world.

In July 2021, Steven Pruzansky, Rabbi Emeritus of Congregation

Bnai Yeshurun in Teaneck, New Jersey, wrote that while "we all believe in romance, love, free choice and human rights," intermarriage had consequences. Pruzansky noted that a quarter of American Jews deem Israel an apartheid state and a full 38 percent say they have "no attachment" to Israel. Rather than coming to the conclusion that this is a result of policies pursued by the US and Israeli governments, or by the different trajectories of American and Israeli Jewish populations and their relationships to liberalism, Pruzansky concluded that this was the result of intermarriage.

"Simply put," he wrote, "polls that purport to measure American Jewish public opinion on any issue invariably count people who identify themselves as 'Jews' to the pollsters, whether or not they are in fact Jews according to Jewish law. They are the products of intermarriage, and they may consider themselves on some level 'ethnic' Jews. They may have a Jewish father or grandfather. They may even have a Jewish mother and a non-Jewish father and still be considered Jews according to Halacha, but their Jewish identity is tenuous and clearly not based on the features of Jewish life that bind all Jews: Torah, mitzvot, love of Israel and the people of Israel, etc."[46]

Pruzansky didn't like what the polls told him about American Jews. So he decided to tell the pollsters, and the public, who the *real* American Jews were. He concluded by calling intermarriage a plague and a second Holocaust.

The debate around intermarriage and how it fits, or doesn't, with halakha and whether it will end, or not, the very idea of Jewishness in America dates back for decades. And the objection to it wasn't always made on policy grounds.

In 1981, my father, in his final year of college, and who had, all his life, been told to marry a Jewish girl, met my mother, Cindy Cardinal, of Hicksville, New York, a young Catholic woman from Long Island, in the dorm building in which they both lived. They began dating.

Eventually, they decided to get married (or, rather, my mother, having spent five years of her life dating this man, told him that they would either get engaged or break up).

My father told his parents. This was a fraught task: My nana had decided to boycott her nieces' weddings because they did not marry Jewish men. And she could have decided to take a similar stand here.

"They were impressed with Mommy," my father told me. When he told them they were thinking of getting engaged (after they were, in fact, already engaged), "they didn't cry, there was no big fight. And I know that my father had a discussion with my mother, and he said, 'I'm not losing Danny.'"

And so they attended my parents' wedding. A few years after that, my mother, on her own, decided to convert to Judaism, thrilling my nana.

"I know for a fact that my mother was, maybe one of the most, you know, happiest moments in her life was when she said she was going to convert. I know that it made her happy, I'm sure because the kids, the grandkids. Then you'd be, you know, you'd be Jewish," my father said. (Actually, according to the Central Conference of American Rabbis, patrilineal Reform Jews who were raised Jewish had been considered Jewish since 1983.)

My mother became pregnant with me and (allegedly) first felt me kick while sitting in a synagogue. I was born and had a naming ceremony (my Hebrew name, Etta Faiga, is after Fannie, my great-grandmother—the one my great-grandfather considered plump, like a tomato). Two years later, my sister did, too.

"When we did the naming in our apartment—when we did the thing for Beth up at Temple Emeth [in Brookline, Massachusetts]—that was everything" to my nana, according to my father.

And then, two years later, my brother was born. We were, at that point, in Toronto, Canada. My brother was born very sick. He spent a week in intensive care. When my father called the mohel everyone recommended to perform the Brit Milah, or circumcision ceremony, the mohel asked for my mother's name. Cardinal, my father said. The mohel stopped him.

She converted, my father assured him. The mohel asked for the name of the rabbi who oversaw the conversion. Her name was Deborah Hirsch.

The mohel said no. He would not perform the ceremony. If she was a woman rabbi, that meant that it wasn't a *real* conversion.

"That's not Jewish enough for you?" my father says he replied. "This is why this religion is dying out."[47]

He hung up the phone. They found a different mohel. They raised us Jewish. But I always remembered that story. That there were, and are, people who would look at my mother, and at me, and say we're not really Jewish.

Every group has its rules that determine who is in and who is out. One young modern Orthodox woman told me that she believed that intermarriage was the worst kind of assimilation because "it means you love them more than you love us."[48] I would not tell her that she is wrong, though I disagree with her, because I am not modern Orthodox. She has signed up to an interpretation of Jewishness that I have not. Or, as Jennifer in Squirrel Hill put it, "People interpret halakha differently. And that's okay." (Her mother converted to Judaism, and she married a man who converted, too. The rabbi at her current Conservative synagogue is very welcoming to interfaith families, but not everybody else has been. Still, she understands, she said. "I don't get offended by it.")[49]

I have even come, as I have gotten older, to have more sympathy for the mohel who turned down my baby brother. I do not agree with him, but his actions were motivated by a certain interpretation of Jewish law and tradition and the role that he wants that interpretation to play in his own life.

But as people reach to intermarriage as a cudgel against those they perceive to be their political opponents, my mother's story has taken on, for me, a new resonance.

As we have seen, the focus on intermarriage in American Jewish life has been carefully cultivated not only, or even primarily, by rabbis and mohels but by philanthropists and social scientists. It has always sat poised somewhere between the religious and the social and political. And it is the latter for which I have no sympathy. To put it another way: disqualifying Jews because of interpretation of religious laws is one thing. Disqualifying them—us—to score a contemporary political point is another.

To say that a person who disagrees with you is not Jewish because they either are intermarried or are the product of intermarriage is to admit that you cannot win an argument on the merits. The people who do this cannot defend their political and socioeconomic preferences or could but do not want to bother with engaging, and so say that those who disagree with them aren't really Jewish. They draw the circle smaller and smaller until only those telling them what they want to hear are standing along with them.

This has real-life consequences for those on the receiving end, dismissed as being faux Jews. But it is only a reflection on those doing the dismissing.

This is to say that it is, of course, hurtful to hear that one's marriage is a second Holocaust from a person lashing out about surveys. But it could be worse. I could be the kind of person who said something like that and considered it an argument.

This is gatekeeping, and the gates are even higher for Jews of color.

One young man I interviewed, whose father is a Japanese man who grew up in Baltimore, told me that, while growing up, he was part of a welcoming congregation and never felt any discrimination from classmates or teachers. Today, though, he hides that he's Jewish, he said, not revealing it as a fact about himself right away, and sometimes not after that. "I don't really know what people would think of me," he said. Jews sometimes "pick" at his identity. He can hear them dancing around the legitimacy of his Jewishness. And so somewhere along the way he decided to keep the fact that he's Jewish more private, even though he says, "That's who I am. I think it's a cool thing. It's something I'm proud of."[50]

Another interviewee, who asked to go by her Hebrew name, Shoshana, told me that, as a Black Jew, "When someone who looks like me walks into my shul and you assume I'm the nanny, that's not a good first impression.

"There's a reason I won't go into certain spaces," she said. Where she lives now, in Virginia, "it's very chill," Shoshana said. But back when she lived in New York City, there was "very fraught inter-community stuff." She was obviously at once frustrated and exasperated with it.

"If I weren't so fiercely determined about it, I would have been missing something important in my life," she said. Instead, "I'm like, no this thing is me, and this thing is important, and you're not going to take this from me."[51]

In a 2019 *Forward* piece, writer and filmmaker Rebecca Pierce, who is Black and Jewish, described the racist backlash facing Jews of color, explaining how some Jews had attempted to paint left-wing Jews of color as only pretending to be Jewish and somehow betraying other Jews by doing Palestinian solidarity work or by embracing their racial identity as well as their Jewishness. "These attacks on Jews of Color and Jews by Choice are not only defamatory and cruel; they are fundamentally harmful to the diversity of the Jewish people," Pierce wrote.[52]

Another interviewee, Max D., is the son of a South African Jewish mother and an Indian father. His father is from New Delhi and is, as Max D. put it, "not a religious man." But he was there with the family, attending Jewish services at their Reform congregation in Austin, Texas.

"I was always made to feel quite welcome at our synagogue and in the Austin Jewish community," he told me. "My Jewish identity has remained very strongly tied to the community that I'm a part of."

Even so, he said, "I think the Austin Jewish community has some work to do."

The Austin Jewish community, he said, is heavily Ashkenazic and "quite white." People who are the products of intermarriage want to be a part of the community, but the community itself is still figuring out "how to make people feel."

"There's a conflict of interests between those of us who represent the more diverse generation and the people who remember it being different," he said.

"I think there is a perception that there is a type of Jew that is a real Jew," he told me. And that manifests itself in unseemly questions and challenges—*Are you half-Jewish? Are you really Jewish? Where are your parents from?*—that essentially serve to make people who are Jewish and want to feel like they are a part of a Jewish community as something other than central to that community.

"I think that the perception that there is a particular type of Jewish person that is the real heir to Judaism, not necessarily even because of practice but because of ethnicity or sense of ethnic identity—I mean, it kind of sucks. It's not a nice thing to confront. I wish it were different."

Max D. also suggested that the policing of ethnic or racial Jewish identity was tied up with narrowing what are politically acceptable positions for Jewish people to have.

"I think people who frame the world in terms of Good Jews and Bad Jews would prefer Evangelical Christians to a mixed-race Reform Jew, because my politics and my identity are inconvenient," he said. "My criticism is inconvenient enough that they'd rather say I'm not a stakeholder than pay attention to my concerns."

Before we ended our interview, Max D. stressed that, overall, his interaction with Jewish spaces was a positive one. "That doesn't mean there aren't problems but I'm still a part of this community," he said, adding, "and it's part of me."[53]

Not every Jewish person of color in the United States has been made to feel less than or left out. One young man, Jamie Yong, whose father's family is from Hong Kong and whose mother's family comprises Jews who came to the United States from what is now Poland, said he wasn't made to feel different growing up. His family was deeply enmeshed in their synagogue—his mother was president of it at one point and very involved beyond that and so knew everyone—and he always felt that their family was accepted there.[54]

But many haven't. According to a study—the largest ever of its kind—released in the summer of 2021 by the Jews of Color Initiative, eight in ten respondents said that they had experienced discrimination in a Jewish setting. Roughly two thirds of the respondents identified as biracial, mixed race, or multiracial and spoke about how they bridged their cultures (one woman hosted a Bollywood-themed Shabbat). The respondents were committed to Jewish life and Jewish practice. But 80 percent said that they'd been discriminated against while pursuing it.[55]

There are people who criticize the term "Jews of color," and in some cases, this is for good reason: for one thing, the experiences of Jews of color are wildly different and varied; for another, some people who are identified by others as Jews of color, like Middle Eastern Jews, may or may not actually consider themselves to be non-white. Ascribing labels to people that they themselves reject undermines efforts to be more sensitive and inclusive to people with different identities and backgrounds.

But in other cases, criticism of the term or its application comes from those who insist that Jews of color are being overcounted[56]—or, in the case of the 2021 study, that the negative experiences of Jews of color are being overcounted. That things are not as bad as the study would suggest.[57]

This criticism rarely comes from people who would be considered Jews of color. Which is to say that it is Jews who see people like themselves in leadership positions in the Jewish world, and who are less likely to be questioned about what they're doing in a shul or where their parents are from or whether they're really Jewish, who are taking it upon themselves to say that everything is fine.

Let us, for a moment, assume that these people are correct and that 80 percent of Jews of color have not experienced discrimination in a Jewish setting. Let us imagine that it was 60 percent, or 40 percent, or 10 percent, or 2 percent.

Wouldn't that still be too high?

There is another irony in all of this, which is that there are some Jews who, for all intents and purposes, go through life in the contemporary United States as white Americans who insist that they are not white.

It is true that race, particularly in the United States, gets mapped onto other identities, because the United States was, from the beginning, a society constructed around race. Jews who identified themselves as ethnically distinct were nevertheless grouped into the basket of "white" by Americans who did not understand the concept of ethnicity beyond white, and by Jewish leaders who understood that whiteness offered legal protection. There have been Jews in this country, since its founding, who upheld white supremacy, or at the very least were complicit in its continuation. This was not because they were Jewish. Being Jewish does not make a person complicit in white supremacy. But actively participating in certain institutions and practices—like white flight—does. Or, as Rebecca Pierce put it in her 2019 *Forward* piece, "we face attempts to write us out of a Jewish community that insists over and over that it's not white, all the while trying to cleanse itself of its black members."

Jews today may say they are not white, but saying so does not absolve us of these elements of American Jewish history. It isn't enough to look at Abraham Joshua Heschel and the Jews who participated in the civil rights movement. That is American Jewish history, but it is only a part of it. In the contemporary context, that means grappling with all of American Jewish history and with the various stances American Jews have chosen to take with respect to white supremacy. It also means that those who say that Jews aren't white only to turn around and malign Jews who do not look white as not really being Jewish are only fooling themselves.

Race is a construct, but it is a construct with lived implications. And there are, in the United States, Jews who go through life as white. This is the majority of American Jews. If they—we—do not wish to be considered as complicit in white supremacy, a good place to start would be by not insisting that we're more Jewish than Jews of color.

One evening, after conducting enough of these interviews, I approached my husband, who is Indian American, to have a conversation. He was sitting on the couch reading a book.

I asked him if, when we have children, people will try to tell them that they're not Jewish.

"Probably," he said.

And what will we do then, I asked him.

He didn't look up from his book.

"We'll tell them they're Jewish," he said.

There are already so many different people—and communities, and publications, and institutions both makeshift and longer lasting— who have gotten tired of waiting to be told that they're Jewish. They are asserting themselves politically, journalistically, and religiously. They have heard that they are Bad Jews before, and they aren't particularly interested in listening to it any longer.

Chapter 9

Pushing Jews

O VER THE PAST DECADE, a few trends have taken hold of American Jewish communities. One is that American Jews, and particularly younger American Jews, have grown more detached from and critical of Israel, and indeed from American Jewish institutional life more generally. Another is that American Jewish institutions, broadly speaking, have not adapted, continuing to pursue the same policies and party lines that they have since the postwar period. Some American Jews felt that Jewish institutions failed them in the Trump era, choosing to stand with Israel instead of with American Jews. And yet another trend is that the political Jewish Right has attempted to cast those who criticize it as either insufficiently Jewish or not Jewish at all.

In light of all of this, some American Jews looked around and decided to do what, contrary to certain narratives, some American Jews have always done: to try to change the way things are in American Jewish life.

• • •

In 2016, some who tuned in for one of the Democratic primary debates were shocked. Senator Bernie Sanders mentioned Palestinian rights.

"I read Secretary Clinton's speech before AIPAC [the American Israel Public Affairs Committee]. I heard virtually no discussion at all about the needs of the Palestinian people," he said. "Of course Israel has a right to defend itself, but long-term, there will never be peace in that region, unless the United States plays a role, an even-handed role, trying to bring people together and recognizing the serious problems that exist among the Palestinian people. . . . There comes a time when, if we pursue justice and peace, we are going to have to say that Netanyahu is not right all of the time."[1]

The comment made headlines and sparked takes and sent pundits chattering. Some wondered how he could say such a thing; others wondered why it had taken so long for someone to say it.

Over the past eight years, there had been something of a realignment of mainstream Jewish American sentiment. Though some latched on to the idea that Obama was hostile to Jews, pointing at his former pastor, Jeremiah Wright, who said the Jews were keeping him from President Barack Obama once he was in office,[2] the reality was that Obama had been close to the Jewish community in Chicago—a broadly more liberal Reform Jewish community[3]—and had butted heads with AIPAC, "America's Pro-Israel Lobby," long seen as the most significant voice on Israel.[4] As such, there was, as one Democratic Jewish analyst explained to me in 2019 on the condition of anonymity, a recognition that the loudest Jewish voices in America were not necessarily representative of mainstream Jewish politics, and certainly not mainstream Jewish Democratic politics, anymore.[5]

The slow appreciation of this growing gulf also coincided with the rise of J Street. Founded in 2007, J Street was set up as an alternative to AIPAC. At first, it was dismissed by detractors as fringe and overly critical of Israel. Over the course of Obama's tenure, the "pro-Israel, pro-peace" group began to be seen as more mainstream.

Jeremy Ben-Ami, president of J Street, grew up like many a liberal American Jew: in a Zionist New York City home.[6] He went to Princeton University followed by New York University Law School. He went to work for President Bill Clinton, did a stint in Israel in communications, came back to the United States, and worked as an adviser to Howard Dean's 2004 presidential campaign.[7]

But then Ben-Ami did what many people with his background would not have done: he cofounded J Street.

J Street has, from the beginning, billed itself as "pro-Israel, pro-peace." It was seen by some as an answer to AIPAC, an insistence that there could be more than one way to speak with a "pro-Israel" voice. "It used to be that there was simply one venue and one conversation," Ben-Ami told me when I interviewed him in his spacious downtown DC office in 2019. "That monolith has been broken."

J Street presented itself as an organization that clearly believed in the importance of the continued existence of a Jewish state but also felt free—and indeed, has as part of its mandate an obligation—to criticize Israel and Israeli policy.[8]

J Street was not—and did not intend to be—a radical organization. It is still staunchly Zionist. Ben-Ami's 2011 book, *A New Voice for Israel*, was criticized in some quarters for saying nothing new.[9]

Despite this, the backlash to J Street, especially in its early days, was furious. In 2011, a hearing on J Street was convened in Israeli parliament. "The conveners of Wednesday's hearing, a hawkish Likud legislator named Danny Danon and a conservative colleague from the centrist Kadima party, Otniel Schneller, wanted to expose J Street for what they believed it to be—a group of self-doubting American Jews more worried about what their neighbors say than what is good for the state of Israel," read a *New York Times* article on the event (Danon would go on to be Israel's ambassador to the United Nations from 2015 to 2020).

The *Times* also quoted Schneller as saying that, despite the fact that J Street calls itself pro-Israel, "This is a dispute between those who care what non-Jews will say and those who believe in being a light unto nations, between the mentality of exile and that of redemption. J Street is not a Zionist organization. It offers love with strings attached. They say, 'We love you only if you behave the way we like.'"[10]

Five years later, in 2016, David Friedman, then president of American Friends of Bet El, wrote, "Are J Street supporters really as bad as kapos? The answer, actually, is no. They are far worse than kapos— Jews who turned in their fellow Jews in the Nazi death camps. The kapos faced extraordinary cruelty and who knows what any of us would have done under those circumstances to save a loved one? But J Street? They are just smug advocates of Israel's destruction delivered from the comfort of their secure American sofas—it's hard to imagine anyone worse."[11] Friedman would go on to be Trump's ambassador to Israel.

The article in which Friedman attacked J Street was actually a response to Peter Beinart, a writer who was thought of at the time as a sort of prototypical liberal Zionist, and who had criticized Friedman's support of Trump. Four years later, in 2020, Beinart declared that he had abandoned the idea of a Jewish state, and instead wanted to imagine a Jewish home in an equal state. Beinart, writing in the *New York Times*, shared that "the dream of a two-state solution that would give Palestinians a country of their own let me hope that I could remain a liberal and a supporter of Jewish statehood at the same time. Events have now extinguished that hope."[12]

Back in 2016, Israeli lawmakers and lobbyists and Likud loyalists could have asked themselves why a "liberal Zionist" organization like

J Street had formed in the first place. Was it really so that Ben-Ami could make himself look good to his neighbors? Would someone go through the trouble of forming an organization and being derided in the Knesset for that? Is the only reason that an American Jew would criticize Israel really because he wanted acceptance from a certain group of other Americans? Was no other reason imaginable? Or was there a broader shift happening in American politics of which this was only a small part? Were American politics changing? Were American Jews coming to be more comfortable criticizing Israel—or, to put it another way, less comfortable not criticizing Israel?

Alternatively, Ben-Ami and company could have decided that they'd been wrong, and conceded that he'd been a Bad Jew, and that actually they were acting in a way that was worse than literal hand-maids to the Nazis, and the appropriate thing would be to apologize and echo Friedman. And everyone who, deep down, agreed with Ben-Ami could have done the same.

Neither of those things happened, and history marched on.

Sanders didn't become the candidate; Clinton did. Sanders supporter Jim Zogby, president of the Arab American Institute and a longtime Palestinian rights advocate, pushed for "an end to occupation and il-legal settlements" statement to be included in the DNC platform. It was left out. Zogby said at the time that Sanders had had a hand in crafting the language regarding occupation. In 2019, he admitted to me that Sanders's support of the issue was something of a surprise: "Two years out, I had no idea that Bernie Sanders would be raising the issue. I don't think Bernie Sanders did."[13]

That same year, during the midterm elections and after two years of Trump, American Jews voted for Democrats at a higher percent-age than they did in the 2016 general election (80 percent did in 2018,[14] compared to roughly 70 percent in 2016).[15] They did this because, for most American Jews, Israel is not the most important political issue. They are more concerned with the United States, the country where they live. Those who closely watched how American Jews voted in the past would have also noted that most voted for Democrats after Obama's Iran nuclear deal, which Netanyahu strongly opposed. Those who watched especially carefully would have known that more American Jews supported the Iran deal than not.[16] And some believe that Trump's embrace of Israel—and specifically, of Netanyahu's Israel—may have made more American Jews than before consider what they themselves had been embracing.

"Just the fact that Trump [who has embraced Netanyahu] has made it possible for more and more Democrats [who] have aligned with this kind of AIPAC-defined, pro-Israel [position] but still would define themselves as liberals—made it possible for them to recognize that what Netanyahu is doing in Israel has nothing to do with liberalism," a progressive Senate staffer told me in 2019, speaking on the condition of anonymity.[17]

Also in 2018, new members were elected to Congress, among them Ilhan Omar of Minnesota, who came to the United States in her childhood as a refugee from Somalia, and Rashida Tlaib of Michigan, whose parents are Palestinian immigrants. Both challenged the bipartisan consensus on Israel. Both have spoken in support of the Boycott, Divestment, Sanctions movement,[18] and Tlaib backs a one-state solution;[19] the vast majority of politicians in Washington champion a two-state solution, even as the possibility that they will see one in their lifetimes seems to slip farther and farther away.

Leftist Jewish action was not limited to Israel-Palestine. A decade ago, Jewish leftists assembled at Zuccotti Park for Occupy Wall Street, planning Yom Kippur services while agitating against capitalism.[20] In 2019, Jews Against ICE (Immigration and Customs Enforcement) protested detention centers around the country, carrying banners that read, NEVER AGAIN. One person showed up with a shofar—the musical instrument traditionally used as part of Jewish religious ceremonies, here repurposed.[21] The protests continued for months. Some wrote op-eds explaining that they were protesting Trump's immigration and deportation policies not despite their Judaism but because of it.[22] For others, protesting ICE was a way to mark the High Holidays.[23] Bend the Arc, a national Jewish organization focused on social change in the United States, stressed the need for solidarity with other peoples and other movements in the fights against white nationalism and for racial justice, civil rights, and immigration reform. Their Twitter is marked by a profile picture reading, "In Defense of Black Life"; in early 2022, it displayed a JEWS FOR THE FREEDOM TO VOTE banner.

There were, and are, some who insist that these American Jews just want to be accepted by cool progressives and sit at the right lunch table, and that, in the process, they are closing their eyes and ears to the more insidious form of antisemitism: antisemitism on the left. Liel Leibovitz, writing in Tablet in 2021, warned, "The Democratic Party, where American Jews long made their home, has embraced intersectional politics—a tactic that, judging by recent election results, is working well for them, if not necessarily for their Jewish members." This, he said, would lead to a disconnect between Jews and their sense of Judaism. "In liberal and non-Orthodox circles, individuals and institutions that devote most of their emotional and spiritual energies to checking off the rapidly shifting boxes erected by the left are going to find themselves with a version of Judaism mostly disconnected from religious observance, and which will therefore attract only the people in the market for progressive activism, with a little bit of cholent thrown in for nostalgia's sake."[24] Meghan McCain, former cohost of The View, has lectured both Senator Chuck Schumer and Senator Bernie Sanders for not taking left-wing antisemitism seriously enough (Schumer and Sanders are both Jewish; McCain is not).[25]

This is part of a broader debate on how, exactly, to define antisemitism. What one considers to be the most dangerous kind of antisemitism depends on what one understands antisemitism to be in the first place, and this is heavily contested. Many established Jewish organizations have pushed for the adoption of the International Holocaust Remembrance Alliance definition to be adopted as policy; the lead writer of what became that definition, Kenneth Stern, said it was never meant to be used as an enforceable code. And since that definition comes with examples of antisemitism that include things like "denying the Jewish people their right to self-determination, for example, by claiming that the existence of a State of Israel is a racist endeavor," some have said that using it as such could have a chilling effect on speech in support of Palestinian rights.[26]

Others prefer the Jerusalem Declaration on Antisemitism, which was written as a kind of alternative to the IHRA definition. Though the Jerusalem Declaration defines "Denying the right of Jews in the State of Israel to exist and flourish, collectively and individually, as Jews, in accordance with the principle of equality" as antisemitic, it also says that "Criticizing or opposing Zionism as a form of nationalism" is not. "It is not antisemitic to support arrangements that accord full equality to all inhabitants 'between the river and the sea,' whether in two states, a binational state, unitary democratic state, federal state, or in whatever form," the declaration says.[27] The Jerusalem Declaration was thus criticized by some as antisemites' preferred definition of antisemitism. David Schraub, assistant law professor of the Lewis and Clark Law School, wrote in *Haaretz* in April 2021 that "the JDA is being understood (again, notwithstanding its own text) by many as a warrant to dismiss *all* antisemitism claims as illicit and illegitimate smears. Those fans of the JDA aren't worried that they will be called out for antisemitism, because they understand the JDA's function is not to call things antisemitic at all."[28] We cannot even agree on what antisemitism is. It is therefore not surprising that there is disagreement as to whether it is more of a problem on the right or the left.

It would be foolish to deny that antisemitism on the left exists.

Antisemitism predates and transcends political affiliation. There is no political corner or crevice where one can hide to avoid antisemitism. I agree with David Nirenberg, dean of the University of Chicago's Divinity School, who told the *New Yorker* in 2020, "I think the real danger is imagining that it is only the other where anti-Judaism is doing its work and thereby not being able to see it in your own affinity group."[29] In 2015, when a student at the University of California, Los Angeles, had her ability to be an unbiased participant in the students' judicial board questioned at length because of her involvement in Jewish life on campus, I winced.[30] When, in 2021, synagogues were vandalized in the United States[31] because Israel was shooting artillery and airstrikes at Gaza in response to Hamas's rockets, which were in turn in response to violence against Palestinians in East Jerusalem,[32] I felt sick (in case it is not clear that the death of civilians made me feel sick, too: it did). When Tamika Mallory, a prominent activist in the fight against police brutality, refused to disavow Louis Farrakhan, who likens Jewish people to cockroaches, I felt sad, even understanding the positive role the Nation of Islam has played in some Black communities.[33]

When, in 2019, Ilhan Omar said that support for Israel was "all about the Benjamins," I groaned, both because I don't think that's factually true—I don't think, for example, that Jewish or Evangelical support is bound up in money, or that the United States sees Israel as a security partner because of wealthy benefactors—and because it was reminiscent of old stereotypes about Jews and money.

But unlike many other politicians who have spoken aloud antisemitic stereotypes, Omar apologized for her remark, saying she was unaware of the trope.[34] She has not apologized for saying that Israel should be accountable to the International Criminal Court, but why should she have to? Saying that Israel should be accountable to an international governing body is not against Jews as Jews, but about the behavior of a country and whether it is living up to the commitments it has agreed to as a member of the international community.

"When the whole backlash against my boss happened," Jeremy Slevin, Omar's communications adviser, who is himself Jewish, told me, "it was so over the top and so clearly layered with anti-Muslim, and frankly racist overtones, it really clarified things."[35]

I asked Slevin what it was like to be Jewish and work in an office that is seemingly constantly accused of antisemitism for calling out what his boss sees as human rights abuses.

"We go out of our way to be consistent in condemning human rights abuses everywhere," he told me. And Omar has indeed been one of the loudest critics against human rights violations by Saudi Arabia,[36] India,[37] China,[38] and Ethiopia,[39] among others. And so, to be accused of picking on Jews, "It's hard to explain because it's like—I was Jewish long before I worked for Ilhan Omar.

"It's almost like you're having your identity stripped."[40]

The other member of Congress most commonly accused of antisemitism on the left is Rashida Tlaib, a daughter of Palestinian immigrants.[41] But of course Tlaib thinks differently about Israel than many members of Congress, and, for that matter, many American Jews. Hers is a different perspective, one that is underrepresented in Washington. And if her detractors really believe that she is wrong when she says that Israel is engaging in apartheid (as Human Rights Watch,[42] Israeli human rights group B'Tselem[43] articles in the Israeli newspaper *Haaretz*,[44] and dozens of rabbinical students in a 2021 letter have also done[45]), they should argue that on the merits, not by calling her an antisemite.

Mostly, though, when I hear that the fixation should be on antisemitism on the left, I recall that there was a reason that American Jewish professionals in the 1960s decided not to focus on the antisemitism within the Nation of Islam and instead to focus on support for civil rights and the struggle against white supremacy—it was to protect, not to surrender, their ability to be Jewish in the United States. Because antisemitism, though unique, is not alone. And in the United States, it works in concert with other hatreds. And fighting it—to my mind, at least—requires a recognition of that.

"In the progressive movement, I always fought for other people's rights," Scott Goodstein, who has worked on a variety of political campaigns and is now based in Washington, DC, told me. As time went on, he said, he became "more and more comfortable standing up for my own rights."

"It took me a while to start standing up for my own personal identity. It was never something that I led with," said Goodstein. His nieces and nephews wouldn't put up with the kind of language he used to tolerate hearing. He doesn't anymore, either.

There are, he said, some antisemites on the left in all movements. But even so, "I think it operates on a different axis than political identity. If Jews are not in the rooms where decisions are made, then shame on us for walking out of the room."

He said that a rabbi by the name of Sharon Brous helped him work through this. He was helping build major parts of the Black Lives Matter digital movement and saw that older American Jews—the ones who were so proud of Jewish participation in the civil rights movement—weren't there with him. And Jewish groups were grappling, and are grappling, with BLM's support for the Palestinian rights movement and criticism of Israel, which, for some, bleeds into the destruction of Jewish lives and security.

Rabbi Brous told him about the Jewish people who were mad at what they felt was language against Jews and claimed, "I'm going to walk away." She explained that this was, for many of them, an excuse. They hadn't been there in the first place, so how could they walk away? And if everybody walked away, who would be there in the movement?

"There's real problems on the left, and it's not just antisemitism. There's racism, homophobia, misogynistic problems.

"The way to deal with it," he told me, "is to show up more."[46]

In 2020, Bernie Sanders, once again a candidate in the Democratic presidential primaries, promised that his administration would redirect US military aid to Israel to the humanitarian situation in Gaza.[47] Sanders, former Housing and Urban Development secretary Julian Castro, and South Bend, Indiana, mayor Pete Buttigieg all showed up to speak at J Street's 2019 conference.[48] Sanders did not attend the AIPAC Summit in 2020. Senator Elizabeth Warren also opted out.[49] Buttgieg did not attend in person but sent a video message, as did former vice president Biden.[50] Mike Bloomberg,[51] the former mayor of New York who spent hundreds of millions on his own election, inviting both fair and antisemitic criticism in the process, spoke in person.

Bloomberg and Sanders were both the sons of Polish Jewish immigrants. But they obviously came to have very different opinions on Israel. And it was the division between Bloomberg and Sanders that was perhaps the most interesting political development for American Jews.

"I am very proud of being Jewish," Sanders said at one early 2020 Democratic debate. "I actually lived in Israel for some months. I happen to believe that right now, sadly, tragically, in Israel, through Bibi Netanyahu, you have a reactionary racist who is now running that country. And I happen to believe that what our foreign policy in the Mideast should be about is absolutely protecting the independence and security of Israel. But you cannot ignore the suffering of the Palestinian people."

It wasn't what he had said in 2016; it went farther.

Bloomberg tried to reply by calling settlements "new communities."

"Settlements," Sanders shot back.[52]

"To characterize AIPAC as a racist platform is offensive, divisive, and dangerous to Israel—America's most important ally in the Middle East—and to Jews," Bloomberg later tweeted. "How can Bernie profess he's the path to unity when he's already managed to polarize a people and a party?"[53]

But Bernie Sanders did not polarize American Jews. American Jews were polarized long before Bernie Sanders's presidential run and will remain so long after his career has concluded.

After the debate, I considered that Americans also saw that night two Jewish men, in front of millions, arguing different Jewish opinions. One had risen up through politics by calling himself a democratic Socialist and aligning himself with workers, and the other had invented information terminals and made billions. They were both, in their way, stereotypically Jewish paths, and, at the same time, together, breaking Jewish American convention. And there they both were, running not for the office of America's Best and Most Correct Jew, but to be president of the United States.

Neither would win.

That Bloomberg dropped out wasn't particularly significant to any portion of America's Jewish community. That isn't to say that he didn't have support from Jewish people—he polled particularly well among some Jewish people in New York, especially compared to Sanders—but he wasn't associated with a shift in a segment of Jewish politics in the same way that Sanders was.

Sanders's campaign suspension was different. As controversial as Sanders was for some American Jews, who tried to dismiss him as not really being Jewish, or as covering for antisemites (such people typically pointed to his employment of certain Muslim surrogates on his campaign),[54] for others—in particular, young Jews who wanted to keep both their identity and politics, both their sense of Jewishness and their sense of righteousness and belief in universal human rights—he was not only a political figure and leader of a movement but someone with whom they could identify.

Sanders, who had his Jewishness questioned over and over again by pundits and in print during and after the 2016 election, spoke about it more openly and more often in the 2020 election. Being Jewish, he said in the winter of 2020, impacted his politics "profoundly."

"When I try to think about the views that I came to hold there are two factors. One I grew up in a family that didn't have a lot of money . . . and the second one is being Jewish."[55]

His Jewish supporters focused on his Jewishness more, too. Jews for Bernie—self-described on Twitter as "A collective of Jews—mensches, yentas, ammahs, bubbes, joons, zaydes, ihitochi, teachers & parents—fighting for our #FirstJewishPresident: Bernie Sanders"—formed and released videos and campaigned. Sanders wrote an essay on antisemitism for *Jewish Currents*[56] and was, in turn, the subject of essays about what it would mean to have a progressive Jewish president.[57]

And then, with more candidates lined up behind Vice President Joe Biden and a global pandemic raging, Sanders announced that he was suspending his campaign. "Please stay in this fight with me," he told his supporters in the spring of 2020. "The struggle continues."

"Bernie Sanders' movement represented the Jewish call to pursue justice. We are heartened knowing that his campaign pushed our politics forward, especially on the need to hold the Israeli government accountable, and inspired everyday people to become the leaders we need," IfNotNow said in a statement after Sanders suspended his campaign.

"Tonight, amidst the coronavirus crisis, many of us will sit down to virtual seders. We refuse to feel hopeless, knowing that our people have always survived dark times with visions of a better future. The success of Bernie's campaign and his ideas is the spark that has ignited a generation. The fight goes on. The movement behind Bernie and his progressive agenda is not going anywhere. Like so many other young people, we remain committed to fighting against the current, failed political leadership and for a more just world for ourselves and our neighbors."[58]

Jews for Bernie put it somewhat more succinctly.

"Thank you, Zayde," the group tweeted. "The rest is up to us."[59]

Biden, meanwhile, embraced J Street's first ever presidential endorsement.

"J Street has been a powerful voice to advance social justice here at home, and to advocate for a two-state solution that advances Middle East Peace. I share with J Street's membership an unyielding dedication to the survival and security of Israel, and an equal commitment to creating a future of peace and opportunity for Israeli and Palestinian children alike. That's what we have to keep working toward—and what I'll do as president with J Street's support," Biden said.

It was a shift in politics over the last four years that led J Street to endorse Biden even though it hadn't endorsed Clinton, Ben-Ami said.

"The alignment in the Democratic Party and the shift of the conversation on our issues allows us to feel really great about lining up behind someone like Joe Biden," Ben-Ami said, adding, "Politics is different in 2020 than it was in 2016, and this issue is no exception. The way the politics in Israel has moved so far right and the way Trump has embraced what's going on there has created a lot more space for Democratic candidates."[60]

By this time, J Street had come to be viewed as sitting squarely in the mainstream. Other organizations, in turn, stood up to do what J Street had once done: to challenge the mainstream. Groups like Jewish Voice for Peace, which has existed since the 1990s, and which, unlike J Street, supports the Boycott, Divestment, Sanctions movement, worried J Street would be a "drag" on progressive energy. If-NotNow, another Jewish American group that was established during the 2014 Gaza conflict, took a similar view. Cofounder Emily Mayer told me over the phone in 2019, "There is just so much of a gap between where the Democratic Party's leadership is on the question of occupation—and Israel-Palestine more broadly—and where the Democratic base is . . . it remains to be seen how much J Street keeps up with that push and with that shift."

"The disavowal of Zionism is a single digit view," Ben-Ami told me when I asked him about groups like Jewish Voice and IfNotNow. "It's minimal, single digits. They're very loud, by the way. And carry a microphone [that's] pretty oversized."[61]

And that was true. But if, in the polyphony of American Jewish voices speaking about Israel, J Street could be heard more loudly, then the voices to its left could be, too.

It would be unfair to say that all the newly established groups in American Jewish life are more critical of Israel. They are not, and it should also be said that some young people have gone in the opposite direction. Some young people, for example, have become more outspoken about antisemitism and defending Zionism on college campuses. In 2021, CNN wrote up a story on two young Jewish people—Julia Jassey and Blake Flayton—who are involved in two such groups (Jassey founded Jewish on Campus, while Flayton is on the board of the New Zionist Congress).[62] And certainly, Orthodox Judaism, the fastest-growing denomination among young people, is not becoming more critical of Israel.[63]

Meanwhile, in Israel, Dov Lipman, who made aliyah from the United States almost twenty years ago and had been a member of the Knesset, said he has noticed a trend: young *olim* (*olim* being new immigrants to Israel). Lipman's group, Yad L'Olim, has set up a department for them—roughly ages eighteen to thirty, single, without families—at the request of some of these younger people. Lipman said that he thought that part of the reason that these individuals picked up and moved by themselves was "anti-Israel sentiment on college campuses, fears about where things are heading in America in the long term.

"If you want to live a Jewish life . . . and see your family continue as proud Jews—that's not only easier in Israel," but increasingly, for these young *olim*, it's seen as "only possible" there, Lipman said.[64]

Further, it is not as though the aforementioned American Jewish groups working in support of Palestinian rights do not stumble and fumble. The Pittsburgh and Austin chapters of IfNotNow disaffiliated. IfNotNow is working to rediscover and redefine what and how it should be several years after its founding.[65]

And IfNotNow is an example of the larger choices facing progressive Jews: What is the progressive Jewish position to take? What are they to do with the reality that many American Jews engaged with these issues have access to travel, higher education, and resources, and are in a privileged position relative to US and Israeli society more broadly? And should these engaged American Jews focus on white supremacy at home, or on Israel, or both? Is it enough to champion an end to the occupation? What about inequality between Jews and Arabs in Israel proper? And what does fighting for change look like? Is it pushing Congress to condition aid to Israel? Is it boycotting Israeli goods? Do you advocate for a single binational state? Should you be anti-Zionist? Non-Zionist? And what does that mean in practice, since Zionism was a movement toward the establishment of a country that now exists, with a government setting policies and a military helping to enact them? At what point are we having a conversation about ourselves, about American Jews, and who we want to be, and not about actual policy in and about Israel-Palestine?

"I get really frustrated and angry and tweet things I delete later. People look for American Jews in America in conversations about Israel," said Klil H. Neori, who lives in upstate New York but was raised in Israel. His family is still there. There is a sense, in these discussions and debates, he said, that Israel and Israelis become an abstraction, which is also "very much true of Palestinians, of course. American organizations [dedicated to] Israel—they're just nothing like Israeli politics."

Neori also pushed back on the idea popular among many liberal Jews that being Jewish is about *tikkun olam*, a responsibility to heal the world. "I find that really troubling. Jews just exist. We're just around. And some of us are assholes," he said. "That's bad because it's bad to be an asshole, but it's not bad because they're Jews being assholes."[66]

And all the debating of how to position oneself politically and policy wise as a Jewish person, and all the writing and books like this one about the responsibilities or lack of responsibilities of American Jews toward Israel can lose sight of that: that we are not the people living and dying because of these policy discussions. To a certain extent, the stakes of the debates on how to be Jewish in the United States are relatively low. They may hurt feelings or may exclude from this synagogue or that school. But that does not compare to the plight of Palestinians, whom even the most well-intended Jewish rally or essay of solidarity can remove from the center.

"I've definitely observed, over time, situations where it's really important to hear critical voices from the American Jewish community who are standing in solidarity with Palestinians or, even if they're not in solidarity with Palestinians, making it clear that dissenting from the normal line of no daylight support from Israel is acceptable," said Yousef Munayyer, a Palestinian American writer and activist.

"Those things are good. They can also be problematic.

"Issues that directly and most severely impact Palestinians are being discussed in debates that completely exclude their voices," Munayyer said. "And the critical voices are other Zionists." When the range is from Jews who won't hear anything bad about Israel to Jews pointing out that Israeli generals also criticized the Israeli government, the range is still too limited and exclusive.

"People playing those roles think they're providing a critical voice," said Munayyer. "To some extent they are. But they're also eating up space that would be used by Palestinian perspective."

In solidarity work, Munayyer said, "you are not the story. You are there to help and support someone else; another group of people you're in solidarity with. That needs to be the foundation of everything you do.

"If you become the story, you're no longer doing solidarity.

"And create more space. Make it a priority to create more space, to broaden the frame of the conversation.

"I think there have been some people who have tried to do that," Munayyer said. "And that should be done more and more."[67]

Jacob Plitman was raised in North Carolina. His parents wanted to make sure that he received a solid Jewish background and education, and so, in the summers, sent him to Camp Judaea, which was, to young Plitman, the "most fun, most amazing place." Faced with the question of whether he was Jewish or American, the camp presented him a clear answer: he was Jewish, and he was a proud Zionist Jew. He worked at the camp and, before college, did a gap year in Israel.

The one rule was that he wasn't allowed to go into the West Bank, which, because he was eighteen, meant that he and a friend decided to go on an adventure in the West Bank. They told the woman they were staying with—they had found her through CouchSurfing, the social networking site that helps travelers find locals with whom to crash—that they were journalists visiting Bethlehem for a Christian newspaper. Their host, an Irish woman deeply sympathetic to the Palestinian cause, had set up meetings for them. And so, for three days, they met organizers. That was when, Plitman told me, he heard the word "Nakba," the Arabic word for the exodus of Palestinians, for the first time. The man from whom he first heard it told him the story of how his brother was killed, of losing his home in Jerusalem in the neighborhood where Plitman was spending his gap year.

"Is he crazy or am I crazy?" Plitman remembers thinking. "Somebody's crazy here."

Plitman went to college. He organized with J Street U, J Street's student arm. He worked for J Street. In May 2016, he went to do service work at a refugee camp in Greece. Trump won. He came back home and started working as a labor organizer. He went to a happy hour because he was "very broke" and wanted to "get drunk for free."

"The guy, Larry Bush, stood up and started telling the story of this magazine" that dated back to 1946.

They had a new grant, and they were looking to hire someone.

"Is this a great idea," he wondered, "or the dumbest idea I've ever had?"

But he decided that, worst case, it would be "a cool place to publish my friends."

"And so I started doing the only thing I knew how to do: organize."[68]

Jewish Currents, the magazine founded for the Jewish Left back in 1946, had fallen largely out of view by the Trump years. But the relaunch of the magazine provided a new space for progressive Jewish discussion. In some cases, articles took the Trump administration to task on its approach to the Israeli-Palestinian conflict,[69] but it also published poetry, criticisms of surveillance of Muslims,[70] and, in one case, a pointed critique of how mainstream Jewish publications treat Jewish writers of color.[71] It also publishes interviews with Palestinians like Basil al-Adraa,[72] a journalist and activist, and essays by Palestinian writers like Kaleem Hawa, who asserted, "For me, as for nearly all Palestinians, the Nakba is a part of an intimate family history. It dispossessed my four grandparents of their homes in Palestine, turning them into refugees. But the Nakba is also a part of our people's present, because it never ended," and demanded the right of return.[73] In the fall of 2021, it ran an interview with poet and activist Mohammed El-Kurd, whose family was among those in the East Jerusalem Sheikh

Jarrah neighborhood fighting to stay in their homes, to which Jewish settlers have laid claim.

"The root of a politics of appeal, or of 'humanization,' is the idea that Palestinians innately are not enough," El-Kurd said, adding, "if we're actually going to humanize Palestinians, it would look like this: The Palestinian is human because they, like everyone else, would slap a person who slapped them."[74]

Jewish Currents was misunderstood by some on the left ("We are not here to revive the Bundt," Plitman said, bemused) and attacked from the right. But that is part of the point.

"Part of what we wanted to build was a place where we could think through what is going on, and what could be done," Plitman said.[75]

I sometimes asked interviewees what they wish had been different about their Jewish upbringing. Was there something they wished they'd gotten more of from Hebrew school? Did they wish they'd gone to a different kind of synagogue, or to have gone more often? Would they have preferred to not have gone to shul at all?

One of the most common answers was that people wished they had been taught more—or, rather, taught differently—about Israel.

"The way we were taught about Israel, and still teach Israel, in Hebrew school was questionable," a young woman named Hannah Kahn who taught Hebrew school for two years told me. "It's led to students coming out of Hebrew school with a certain notion of Israel that I don't think is fair or good. Kids can handle and understand so much more than we give them credit for."

She added that she thought that labeling all criticism of Israel as antisemitic was both wrong and, from a pragmatic standpoint, misguided.

"I worked in politics . . . if you want people to come over on your side, you have to be strategic and thoughtful about convincing people," she said. She referenced the Ben & Jerry's brouhaha of 2021, in which the ice cream company, founded by two Jewish men, announced it would no longer be selling its products in the West Bank. "This is an American ice cream company founded by two liberal Jews . . . the whole thing is the epitome of American Jewishness to me," Kahn said. "And yet they are being demonized and called antisemitic. It blew my mind."[76]

Ironically, one person who agreed that American Jews were poorly educated about Israel was Elad Strohmayer, spokesperson at the Israeli embassy in Washington, DC. The problem, he said, is that "when they're growing up, and they go to Jewish day school, and they hear from their parents how amazing Israel is and it's a paradise and maybe they go even on a Birthright trip . . . and then you get to college campus. . . . And they all of a sudden discover that Israel is not perfect. . . . They hear criticism about Israel, and they never got the tools to deal with it. . . . But I think the key to solve that problem is actually to better educate . . . we should sell the real Israel, the beauty of it and the challenges of it."[77]

This still assumes, though, that Israel should be a central point of American Jews' Jewish education. And that belief—that Israel should be significant to American Jews—is itself being questioned.

"You're raised to think that Israel's this really important place, and, no matter what, we should be defending it," said Jesse Chase-Lubitz, a journalist whose brother, after spending three years in Palestine, "came back and converted the family away from that part of Judaism." She has received pushback, she said, for that change of mind and heart.

"I was once with a group of people and someone called me a self-hating Jew, which was such a hateful term to hear," she said. "Would I be this passionate about this issue if I hated Judaism? Would I hold Judaism to such a high standard if I hated it?"[78]

Logan Bayroff, who now works as vice president of communications at J Street, said that, while he was raised with a "standard

Zionist education," late in high school he started to question some of it. "We're being presented with a pretty slanted view," he recalled thinking. "It actually isn't even aware of how ignorant a lot of it is about the Palestinian perspective."[79]

They aren't alone. Younger American Jews as a whole see Israel as less important than their elders did and do. According to the Pew 2020 survey on American Jews, only 48 percent of American Jews between the ages of eighteen and twenty-nine feel very or somewhat attached to Israel, compared to 67 percent of those ages sixty-five and up. And only 35 percent of those ages eighteen to twenty-nine believe that caring about Israel is essential to being Jewish, compared to 51 percent of those fifty or older.[80] Perhaps this is because, for many of them, familial trauma, in the form of either immigration from persecution or the Holocaust, is farther removed. And perhaps it is also in part because what they are taught growing up is increasingly at odds with what they see out in the world. Israeli settlers continue to try to take over the occupied West Bank.[81] Netanyahu may be out of office, but his successor, Naftali Bennett, insisted that Palestinian statehood would be a mistake.[82] I spoke to one lawyer in Israel, an American who had made aliyah over twenty years ago, who believed that, if Israel turned over the West Bank to Palestinians, it would become another Gaza, and Israel couldn't have that. He asked me to consider the country's difficult neighbors.[83] He lives in Israel, and I do not. But many young American Jews have considered the difficult neighbors and nevertheless decided that they cannot accept the permanent occupation of a people as a logical conclusion.

I told Plitman this. That there are many people who grew up and looked back at what they once learned about Israel and sounded heartbroken.

"So much of what organizing is," he said in response, "is getting people to feel the problems they think are personal problems are actually political problems." The disappointment, the anger, the heartbreak that many American Jews, and particularly young American Jews, articulate with respect to Israel—these, to Plitman, are all reflections of a political problem.

I asked him for his thoughts on the term "Bad Jew."

"For me," he said, "the argument used to be much more hurtful to me when I cared more about the concept of Jewish unity or Jewish community as such.

"But there is no Jewish community," he said. As a monolithic or hegemonic entity, he told me, the Jewish community does not exist. His life experience and that of the traditionally Orthodox, he said, "has almost nothing in common. I don't think they're not Jews. It just has nothing to do with the way I am Jewish."

And then he said something that some American Jews would find unacceptable. "I think that is just fine."

There are some people, he said, who are invested in the idea of Jewish unity. But if that stops making sense or having meaning, "and the differences become as interesting or important as the similarities, then anyone who stands up and says, 'I speak for you,' or a military that says, 'I kill for you'—who's 'you'? Who are you talking to?"[84]

There may not be a Jewish community—one hegemonic, monolithic concept capable of speaking in one voice—but there are Jewish communities. The question is which communities welcome whom, and who feels at home where.

Many American Jews, particularly younger American Jews, felt themselves disconnected from or insufficiently represented by mainstream Jewish institutions.

"I sort of got radicalized by watching the failure of the American Jewish establishment through the 2000s and early 2010s," Jordan Fraade, a Brooklyn-based transportation policy consultant, told me. "The fact that the Jewish establishment has decided to do PR for Israel no matter what—if you're young enough to have only experienced the state of Israel in its modern form, where its political identity is very right-wing and religious, it's hard to feel that they have your best interests in mind, or are up to the task of defending a more pluralistic version of Jewish identity."[85]

And it isn't only around Israel that people are dissatisfied. "Jewish institutional life had a lot of assumptions about what Jewish people were interested in and responded in a lot of ways by saying, 'What if we made our content less Jewish, and that way it'll be more engaging,'" said Jaz Twersky, a rabbinical student. "But if I just wanted to learn about pop music, I wouldn't go to my rabbi for it."

And so some American Jews have tried to shape communities of their own, carving out spaces in opposition to those that alienate, discriminate, and make Jewish education seem out of reach.

SVARA, a "traditionally radical" yeshiva, aims to make Talmud accessible to everyone, and to give students the tools to study it through a queer lens. And Ammud offers Jewish education on things like conversational Hebrew and Torah study for and by Jews of color.

Those involved don't always feel like their community-building is recognized by American Jewish institutions.

"I do really believe that SVARA is filling a niche and meeting a need, and that the Jewish institutional community really owes it for engaging people in a way that they were really not being engaged before," Twersky said. "I absolutely think there's a reluctance to acknowledge work that institutions like SVARA and Ammud are doing, and work other people are doing because they're not doing it through traditional institutions."[86]

Several queer American Jews with whom I spoke told me that they were seeking out spaces that collapsed the distinction between, say, Reform and Orthodox Judaism, where they could have a deeper dive into the texts and hem more closely to traditional services but also not need to deal with gendered seating.

Meli Sameh, who lives in Seattle and described themself as "pretty headfirst in this whole Judaism for weirdos thing," told me that they were "longing for a traditionally practicing Jewish community but not one with a lot of gender expectations." Sameh now sends out a weekly newsletter called Weird Jewish Digest and is trying to build a traditionally accessible Jewish community both in person and online.

"I do feel like it's important to just be like—yeah, there's a lot of Jews that don't fit in the American Jewish community," Sameh said. But "you don't get to make me less Jewish any more than I would get to make you not Jewish."[87]

When I put to Mordechai Lightstone, a Chabad rabbi, the belief some have expressed that Jews in America are splitting between the Orthodox and non-Orthodox Jews, two extremes moving apart and refusing to intersect, he pushed back. Jews today, he said, are in fact far more intersectional in their Judaism, picking from the various diverse options available to them. "If some people are going to synagogue less, that's not a good thing. I want people to go to synagogue. It should be meaningful." But if it's not, he said, then there should be other models.

"American Jews—they can do their social justice volunteering with one group, and have a meal with a Chabad rabbi, expressing their complex Jewish identities through a myriad of ways," Lightstone said. "I think those walls and divisions are beginning to fall in America."[88]

Ike Swetlitz, who now lives in Albuquerque, spoke of hosting Jewish gatherings in Boston, in Washington, DC, and now in New Mexico. Swetlitz pushed back against the idea that Jewish institutions are collapsing. "A lot of people who I've met or connected with in these communities are looking for some sense of connection and structure and I feel like—I guess I chafe a little bit at hearing that we don't want institutions," he said. "We still want organization, commitment, and community.

"A minyan of twenty- and thirtysomethings—is that an institution? It feels like an institution to me," he said.[89]

At the same time, some are questioning the *Protestant, Catholic, Jew* model of understanding Jewishness and Judaism, in which Jews are just like Christians who go to another house of worship. While the Pew survey divides between "Jews of religion" and "Jews of no religion," there are those rejecting that as a framework for understanding Jewishness. In May 2021, when Pew tweeted out that 26 percent of US Jews believe in the God of the Bible, compared to 80 percent of US Christians, Rabbi Danya Ruttenberg interjected, "Howzabout the God of Rambam? Buber? Heschel? The Ein Sof? Ask a Christian-phrased question, get a Christian-phrased answer. (And yes, Jewish atheists: real and valid.)"[90]

"As I have gotten older, whether or not there is an actual metaphysical being who can enact its will on the universe has not been particularly central to my sense of Judaism, nor has it really interfered with my ability to talk about the divine," said Evan Mintz, a Houston-based journalist.

"I'm most certainly not spiritual. I don't want to make fun of other people's spirituality. But I'm religious," said Mintz, a member of a Conservative synagogue. "I like religion, I like ceremonies. They are important and have value." That, he said, is what lays the foundation for Judaism, and for Jewishness.[91]

"I'm not religiously observant at all," Eli Valley told the website Popula in 2019. "But it's not only cultural either! Judaism is such a multi-faceted identity, and expression of one's essence, that it encompasses both culture and values."[92]

Along with this, what constitutes Jewish religious observance itself is being rethought. In *Beyond the Synagogue: Jewish Nostalgia as Religious Practice*, Rachel B. Gross explores how visiting museums, eating certain foods, or tracing family ancestry can actually be understood as religious, as opposed to exclusively cultural behavior, contesting the idea that American Jewish religious practice is declining. "As a mitzvah," Gross writes, "nostalgia for immigrant pasts accommodates the diverse religious needs of American Jews, providing meaning on personal, familial, communal, and institutional levels. . . . Nostalgic practices make American Jews feel a connection to the past that creates community in the present and imagines a particular kind of future."

Gross invites her reader to recognize practices of American Jewish nostalgia as religious, an act that would let us see "the variety of ways that Americans find constructive meaning in their lives beyond traditional denominational structures, in institutions and thorough practices generally considered secular. Redefining American Jewish religion expands where we see Americans finding meaning and which institutions we recognize as most powerful in their lives."[93]

Multiple people spoke to me of the importance that music has in their Jewish lives.

"My status as a creative person is tied up very closely with my identity as a Jew," said Anthony Russell, who, since formally converting, has become a professional Yiddish performer. "It's basically come to pass that when I sit down, and I decide I'm going to create something as an expression of myself, Jewishness is the easiest place for me to go.

"Even creative explorations of my own Blackness are often informed by references to Jewishness and vice versa. As an artist, I think of myself as being a site of *eynigkayt*—singularity—of combined Black and Jewish expression," he added.[94]

Another manifestation of this was the newfound enthusiasm for Yiddish classes. In the summer of 2020, enrollment in Yiddish courses at YIVO Institute for Jewish Research was up 60 percent. Workers Circle, where I take online classes, saw a 65 percent increase from the previous summer.[95] Part of this is undoubtedly because we were in a global pandemic and more people were on Zoom. But there is more to it than that. After all, there are many things one can do via Zoom that are not learning Yiddish.

"In every generation there is news of a Yiddish revival and also news of the death or decline of Yiddish," my Yiddish teacher, Mikhl Yashinsky, known by me as Lerer Mikhl (teacher Mikhl), told me. Yiddish and Yiddish speakers and students have always been there.

Still, he conceded that there is excitement around Yiddish now, and that it is manifesting in different ways. There was a production of *Fiddler on the Roof* in Yiddish (Lerer Mikhl was in it). There are TV shows in Yiddish, like Netflix's *Unorthodox*. There are also young Hasidic people who grew up speaking Yiddish, some of whom helped develop the Yiddish program on the language learning app Duolingo. "That's all meaningful," Lerer Mikhl said. "Part of a language truly being alive is people creating in the language and using it. Not just appreciating it, but creating in it and speaking it."

There are different reasons that people who didn't grow up speaking Yiddish, or who may need to seek out opportunities to speak it in day-to-day life, are drawn to it, Lerer Mikhl said.

"Certainly, there are those who see it as an alternative to forming Jewish identity through Israel or religion. That's not the case for everyone. It's not the case for me. Yiddish is one way I connect, but I also am, from my early days, passionate about Torah and the Land of Israel and all of it. I think there are a lot of people who see it not as an alternative but as a wonderful facet."

Lerer Mikhl also spoke of his connection to his family and feeling close to them through Yiddish. "All of my grandparents spoke Yiddish, and their parents, and their parents," he said. "I always feel that. And somehow feel their voices and their influence on my life. And that's something that brings me fulfillment."[96]

I thought of myself, beginning to learn Yiddish in the pandemic, and of my great-grandparents. They would be bemused that I was spending my time learning this language, I thought then. But as I kept learning Yiddish, I thought that maybe they'd be happy, too. Because my great-grandfather came to this country and now his great-granddaughter has enough security—and curiosity about him and his life—to learn the language of the world he left behind. I think of that, and of him and of me and of my grandfather and father in between us. I think of my father texting me that being Jewish is in your heart and in your history. I think of that, and of this language and culture that is different than it once was but is still somehow alive, and I feel as close as I will probably ever feel to the God I'm not sure I believe in.

At the Museum of the Jewish People in Tel Aviv, I read a quote by George Steiner emblazoned on the wall: "The Jew has his anchorage not in place but in time, in his highly developed sense of history as personal context." I took a photo of it and moved along.

Just as important, there are American Jews shaking the table to challenge who gets to count as an American Jew.

High-profile Jews of color have challenged our conceptions of how we interact with, treat, and make assumptions about one another. Rabbi Angela Buchdahl, the first Asian American person to be ordained as a rabbi or cantor in North America, in her synagogue's Yom Kippur service of 2020 (or, if you prefer, of 5781), urged listeners to think of Jews not as a race but as a family. "Jewish peoplehood is powerful and real," Buchdahl said. "But too often we misunderstand what that means.

"Jewish peoplehood is not about a pure Jewish bloodline, or survival of one strand of DNA. We Jews have never been just one color, or one cluster of chromosomes," she said. She spoke of how a Black congregant told her that a security guard trailed him the first time he came to the synagogue and that he faced questions from other congregants every week. "The impact of this is undeniable. Jews of color experience racism in our community.

"If we don't fundamentally rethink our tribal, racial notion of Jewish peoplehood," the rabbi said, "we are sunk. . . . Instead, think of us as family. You can become family by birth, adoption, and choice."[97]

"Whatever people believe about rabbis, I have just shattered that,"

Rabbi Sandra Lawson told a news crew in the winter of 2021. "As a rabbi who was denied positions based on race, and I know it was based on race, I now get to be the person who makes sure that never happens again." Lawson was the rabbi on campus at Elon University in North Carolina. Her students, she said, will not grow up thinking that women, Black people, and queer people can't be rabbis. They already had one who was all three: her.[98]

She became the director of Reconstructing Judaism. She saw change slowly happening and decided that that wasn't good enough. There was more to be done. So she went to do it.

In 2020, following the murders of Ahmaud Arbery and George Floyd and the police killing of Breonna Taylor, "Many in the Jewish community who are white wanted to know how we got here," Lawson told me. And as a Black rabbi with a law enforcement background (she was in the military before becoming a rabbi), "I was in a very unique position to talk to Jewish leaders and congregations and Hillel [chapters]. I got a lot of speaking engagements or requests for help.

"I was getting asked to come to organizations that saw themselves as very progressive," she said. But when she went to their websites, she saw that they didn't actually have people of color on staff. And she noticed that Reconstructing Judaism didn't, either. So she reached out to senior leadership, applied for a grant to fund her position, and joined. She is, she told me, "committed to helping our movement, which has deep aspirational goals of moving toward anti-racism." And she was going to push the movement toward those goals.

Fifteen percent of Jews under thirty in the United States, she told me, are Jews of color. "And it's just going to keep growing. To continue to keep acting like Jews of color don't exist or are some kind of rarity . . . you're just not paying attention."[99]

Still, the onus shouldn't be, and can't be, on Jews of color to make sure that they're at the proverbial table for discussions.

"If there is an effort coming from an American Jewish space and some aspect of this that's touching on Blackness, Black people should be in the room, Black people should be involved," said Russell, the Black Jewish Yiddish performer.[100] In 2020, he helped translate "Black Lives Matter" into Yiddish along with a group of Yiddishists, an effort that respected both the language and acknowledged that the word for "black" in Yiddish, "shvartse," has a history of being used as a racial slur.

Russell was brought in by Jonah Boyarin, a Yiddish translator and liaison to Jewish communities at the New York City Commission on Human Rights. Boyarin had received a message from a Yiddish student of his in the early days of the protests that responded to the murder of George Floyd. She wanted to know how to say "Black Lives Matter" in Yiddish. "I realized that someone like me, for whom Yiddish is a living, breathing thing that I read, write, and speak, I needed to be able to say, 'Black Lives Matter' in Yiddish. I needed to be able to say it, support it, and talk about it, to write about it, to work on it." Boyarin reached out to a network of Yiddishists and reached out to Russell.[101]

"Oftentimes, that's not the case," Russell said. Black people, including Black Jews, are left out of such conversations. "There's no excuse for that."

Jews of color outside of rabbinical spaces have also sought out space within Jewish communities. When I asked Shoshana from Virginia what her ideal American Jewish community would look like, she told me about a convening of Jews of color that she attended five years ago. People came from all over the world, and a "sizable chunk were queer and/or trans."

"It was amazing," she said. "It was very warm and very welcoming." Anyone who identified as a person of color was welcome—"We weren't gatekeeping"—and attendees ran the gamut from secular to "borderline ultra-Orthodox."

"That was the kind of environment I would love to create on a more consistent basis," Shoshana said.[102]

These changes are being made on the page as well as in person. Michael W. Twitty, who is Black and Jewish and the author of a book on African American cuisine in the Old South, turned his attention to the intersection of American Jewish and African American cooking in *Koshersoul* (Twitty also tweets under the handle "KosherSoul," embracing and celebrating both Jewishness and Blackness with every tweet).

Hey Alma, the online Jewish feminist publication, regularly runs articles featuring Jewish stories often left out of the mainstream narratives about American Jews. They ran an essay by an Asian American Jewish person on anti-Asian violence.[103] They publish conversion stories.[104] They share people's thoughts on reclaiming their Sephardic heritage.[105]

"Younger, progressive generations have placed more emphasis on your many layers of identity," Molly Tolsky, the site's founder and editor, told me. They are "wanting to highlight that and wanting to connect with others over those identities." Hey Alma has also opened her eyes, Tolsky said, to all the different ways there are to grow up as a Jewish person, come to Judaism later in life, and to be Jewish.

"I see a lot of older generations of people in the organized Jewish world ready to write off the younger generations," Tolsky said. But that's a mistake. "People really care. It's just about finding their own way to it that actually feels authentic to them."[106]

Still, these various efforts are not without their pitfalls and problems, either.

"I would describe the current moment as a tokenizing moment," said Devin Naar, the Isaac Alhadeff Professor in Sephardic Studies at the University of Washington. "They realize that there's a problem and they're willing to make a little space. What has not happened yet is grappling with the main narrative—the main story." Adding caucus groups and offshoots for people of color and Sephardic and Mizrahi Jews may be well-intentioned, but it can also serve to reestablish the dominance of white (or, if you prefer, "white") Ashkenazic Jews in the United States.

Naar spoke about the disconnect, growing up, between Jewishness as he saw it in his home and Jewishness as it was represented all around him. It was only very recently, on an episode of *Schitt's Creek*, that he heard a joke about Sephardic Jews specifically (Eugene Levy, who plays the father on the beloved sitcom, is Sephardic). Naar recalled how shocked he was to hear it. "On television! It was like a major moment." So, too, did he speak of the ways in which American Jews have refused to see his family's story as part of their own, with some going so far as to insist that he couldn't have had family members from Greece who were at Auschwitz, because they were from Greece. In actual fact, 87 percent of Greece's Jews were killed in the Holocaust, and the city from which Naar's family comes, Salonica (Thessaloniki), was once home to the largest Sephardic Jewish community in the world. Ninety-six percent of that community was lost to the Holocaust.[107]

"What has not happened yet—nobody has thrown down a big sledgehammer and said, 'The system is not working.'"[108]

It makes one wonder—it makes me wonder, anyway—whether, when people talk about the end of American Jewishness, or the disconnectedness of American Jews, they have bothered speaking to the people with whom they purport to want to connect. There may not be one recognizable way to live a Jewish life in the United States, but there are many. People are finding and forging them all the time.

I can understand, of course, why all of this would be resisted.

Change often is. But I have come to believe that, sometimes, when a person—or a group of people, or an institution, or an establishment, or a mainstream—tries to hold on to something so tightly, because it is precious, or because it is how it's believed to have always been, they end up holding so tightly that the thing itself cracks apart. And then the only thing to do is to rebuild again. Or, failing that, find the piece you feel is worth holding on to.

Tomer Shani, an Israeli lawyer who now lives in New York City, told me that he doesn't feel primarily Jewish. He's first "Tel Avivi," and then a liberal Israeli, and then, and only then, a Jew. There are so many differences between an Israeli Jew and an American Jew, he said. Israeli Jews serve in the military; American Jews (or the vast majority of American Jews, anyway) don't. "Life in Israel is much harder than life in America."

And then he said, "There's an inherent difference between living as a majority and a minority.

"Our Jewish identity is much more natural to us. We're not as paranoid about it as American Jews."[109]

There is a certain paranoia to American Jewishness. There are seemingly regularly occurring essays about whether American Jews have lost what makes us distinct. There are philanthropists up at night worrying that we're going to intermarry ourselves out of existence. There are people who think American Jewish identity is shallow, or, worse, that it's irrelevant. Over and over again in Israel, I heard from people, and particularly from Americans who moved to Israel, that a nice thing about living there was they didn't have to think about how to be Jewish. That doesn't mean that Jewish identity isn't contested in Israel. One need only look at the fights over which denominations get to pray at or near the Western Wall to see that it is. What they meant was that Jewishness was all around them. To be Jewish, they could just be.

And I thought, listening to them, that I was sure that there is something nice about that. But I couldn't, and can't, really relate to the sentiment. To me, trying to figure out what it means to be Jewish, and what is meaningful to me about being Jewish, and finding and holding on to family and community, and making mistakes and changing my mind and trying again: that's my favorite part about being an American Jew.

CONCLUSION

O N A CRISP FALL day in 2021, I stepped, for the first time, into the synagogue that I had joined a year earlier.

I normally say that my husband and I joined a synagogue (during a pandemic, of all times) because I knew that I wanted to raise our future children to be Jewish, and to know that this tradition and history and faith is theirs to choose or reject, and because, from the first days of our marriage, I wanted to consciously incorporate Jewishness and Judaism into our lives together. And all of that is true. But the other reason is that, in doing early interviews for this book, I found myself feeling jealous of people who talked about their synagogues, and the community they found there, and the ways in which they explored their Jewishness there. And it occurred to me that I was an adult, and if I felt that this was something I wanted to do, and my husband was fine with doing it, then we could. It didn't matter if my family didn't belong to one growing up, or if I didn't think of myself as a religious person. There were other Jewish spaces where, in the past, I'd felt less than welcome, or that, in the present, wouldn't embrace my family, but that was about them, not about me. And I could try to find a place that wasn't like that.

I was at the synagogue to interview the rabbi, Daniel Zemel. I wanted to ask him about himself, and about how he thought of Jewishness and Judaism, but mostly I wanted to ask him about tiles.

A few months before our meeting, Rabbi Zemel had given a sermon about thinking of Judaism and Jewishness as a mosaic.

If you only thought of Jewishness as one thing—one song, or one experience, or one idea—you only had one tile in your mosaic. And so of course you would hold on to it. But you would also have a very limited picture. It's as your Jewishness grows and expands and is challenged and imbued with new meaning that you get new tiles and can move the ones that aren't serving you out of the picture. You can rearrange and rearrange and create a new picture.

We sat out behind the building, and I asked him how he came to think of Jewishness in this way. The rabbi explained that this idea came from his reading of an article by Lewis Newman, Professor Emeritus of Carleton College, on studying halakha. Newman compared it to case law, which he in turn compared to a book with different chapters, each written by a different author. There had to be some connectivity. You can't deviate completely. But a new author will indeed introduce changes.

That eventually brought Zemel to a mosaic. A mosaic needs to be in conversation with the larger space around it. And the picture needs to make some sense. You could not, for example, put a cornstalk coming out of the ocean, Zemel said. But within that larger space and sense, the mosaic will evolve.

I asked him what he thought about the future of American Jewishness. "There is a possibility for a very bright future for American Jewish life," he said. "I also think we'll struggle to get there." He worried about national institutions' openness to education. He worried about oligarchic tendencies in Jewish funding. He worried about rabbinic education. And he worried about nostalgia. "We have to learn to look ahead," he said. "There is so much to look forward to."

I thought about the Jews who, in the late nineteenth century, brought over Rabbi Jacob Joseph because they were worried that their Judaism was falling apart, only to strip him of his power because they didn't want to follow one man's rules. And I thought about the Jews who, in the postwar period, looked back to Eastern Europe and the Lower East Side as a more authentically Jewish time and place while moving out to the suburbs and building their synagogues and raising their children. And I thought about Jews today, who look back to that postwar period with its synagogue membership and clear narrative on Israel and worry about what we're losing now.

And I thought about my own family—about Solomon and Fannie, and my grandpa and nana, and my parents, and me—and how different our lives were. How many times the mosaic changed. And how, despite that, or maybe because of it, I can still see the picture.

I asked Zemel what he thought when he heard the term "Bad Jew."

A Bad Jew, he said, "is the person who does not understand that the Torah begins with Adam." With our humanity.[1]

A couple of days later, I sat before my Zoom and listened to his Shabbat sermon. He was speaking about false idols, and how they surround us, and that what's really important is recognizing that we are all made in God's image and seeing humanity in one another. I wondered whether some of these narratives we tell ourselves about who we are, and some of the tenets mainstream Jewish institutions cling to, and the rules they assert need to be followed—if that did not constitute the worship of false idols. I wondered whether some might not look at me and think I was doing that very thing.

But on that day at the synagogue, I turned off my little recorder and walked to the front of the building. Rabbi Zemel asked which of the interviews I had conducted for this book had been the most interesting. And I told the truth, which was that I was surprised to discover that the one that I found most interesting was my mother. I told him it was because of the spirited way in which she responded to the idea that she and her family weren't Jewish enough ("I am a serious Jew in the sense that I take the morality of Judaism really seriously," she said, indignant over the phone. "If you're going to say stuff like that— where's your Judaism?")

But the real reason, which I did not have the words for in the synagogue parking lot, was that she said, "I think religion is something you should think about and internalize. That's the whole point of it."[2] And the reason I found that so interesting was that, when she said it, I realized that I'd ended up back where she and my dad had started. Thinking about being Jewish and trying my best to understand what it means, and then to go out and live accordingly.

I walked away from the synagogue. And I thought about how it—and that last year, really—had given me what I wanted from Jewishness: a sense of connectedness and tradition and meaning, and also a sense of discomfort and challenge. I thought of how I wanted every American Jew to have that, if that was indeed what they wanted. And I didn't want them to have it only in synagogues, the memberships of which are dwindling. I wanted them to find it wherever they could, and wherever was meaningful to them, in their American Jewish lives.

At one point, Rabbi Zemel told me that it sounded like I understood everything. And I said that no, actually, the more I read and wrote and thought about being Jewish, the less I understood of it. I thought, but did not say, that that was all right. I thought, but did not say, that there was plenty on which he and I disagreed. But that I thought that maybe that was part of the point. And that I at least had this mosaic. That I could keep trying, over and over, to get the picture right. I could arrange the tiles so a cornstalk grew out of the ocean. And then I could rearrange them again, water washing away earth, and then, like Noah after the flood, build on what had been and try once more to make something beautiful.

ACKNOWLEDGMENTS

THIS IS A BOOK about Jewish identities and how we describe, process, live, and evolve with them. I want first to thank everyone who got on the phone or Zoom or met in person to try to describe what Jewish identity meant to them.

I also want to thank all of the Jewish studies scholars and writers who I interviewed or quoted in this book. It is incredibly humbling to be in conversation with all of you. This book would not have been possible without the work you have done.

While we are speaking about what would not have been possible: It is not an exaggeration to say that none of this would have happened without my agent, Noah Ballard of Verve. Thank you for everything, Noah.

Thank you, Rebecca Raskin, formerly (no!) of Harper, for wanting to work with me for a second book. I wish I could be edited by you forever, but am so happy for and proud of you for exploring this new career.

Thank you also to Sarah Ried, currently (yay!) of Harper. It is a wild thing to have an editor leave in the middle of writing a book, but even more wild to then find yourself working with a new incredible editor who somehow understands your project even though it wasn't hers. Thank you for your graciousness, your patience, your understanding, and your edits. I'm so grateful to have gotten to work with you.

Joy Asico, thank you for taking my photo and humoring me by including my dog.

Liz Walsh, you are incredible. Thank you for checking my facts.

To everyone at the *New Statesman*, thank you for being the best. Thanks especially to India Bourke and now Alix Kroeger and Megan Gibson for dealing with me day to day; Ido Vock for trolling me; Jeremy Cliffe for seeing that I could be part of this team; and Jason Cowley for hiring me and also never asking why so much of what I file is on American Jews.

Veronica Mooney, Akbar Shahid Ahmed, Zack Nolan, Lauren Fish, Vera Bergengruen, Chris Geidner, Ruby Mellen, Boer Deng, Steve Lucas, Jacob Brogan, Bethany Allen-Ebrahimian—thank you for being there for me here in DC.

Brandon Tensley and Fuzz Hogan—thank you for being here for me here in DC but also in our semiprofessional, semi-petty group chat. Brandon, I must also thank you, as well as Rhaina Cohen and Dr. Marisa G. Franco, for time and feedback on certain parts of this book in our writing group.

Katie David—thank you for always providing a gut check.

Diplomaslack—thanks for being there for me online. I worked out so much of this book chatting between scheming and stabs.

Neel Patel aka NeelKat: I told you I'd thank you in this book. Tell Nyari she can rip the book jacket off if she wants.

Ed Delman—thank you for being, and staying, my fellow Jewish friend from Manhasset. Judah, when you can read this I want you to know: you were a very cute baby. Matt Deinhardt—once again, you are being named so that you can't say you weren't.

Dr. Arlette Sanders took it upon herself to enrich my high school education, and then, when that was done, my life. Thank you, Dr. Sanders.

Thanks to Temple Micah in Washington, DC, for always reminding us that "reform" is a verb.

To my mother-in-law, Pat Bhatiya, who sent us an interfaith card during our first holiday season as a married couple—thank you for being so wonderful.

My siblings, Elizabeth and Nicholas Tamkin, did nothing particularly helpful but they are my siblings, so thanks to them.

Cindy Cardinal, my mom, quite literally changed my life by deciding to convert to Judaism and raise us as Jewish kids. She was often not treated warmly by members of the Jewish community. But I'm not writing this for them. I'm writing this for her. Mommy: I'm sorry anybody ever made you feel less than. I hope you know that, to me, you are the best Jew a person can be, because you are the best person a person can be.

To Dan Tamkin, my dad: Thank you for reading and editing this book for me. Thank you for debating it with me. Thank you for telling me, when I worried I wasn't Jewish enough, that I was a good author for this book. And I'm sorry Grandpa and Nana aren't here to read this, because if they were, I'd thank them for passing Jewishness on to you so that you could pass it on to me.

To Shiloh: You can't read this, because you're a dog, but you're the best girl.

And finally, to my husband, Neil Bhatiya: What is there to say? You're not Jewish but you agreed to be married by a rabbi and raise our future kids Jewish, and you make us watch *Fiddler on the Roof.* You bought us our first mezuzah. You agreed to join a synagogue and do Shabbat dinner every Friday. You cook when we host Passover. You traipse out to see Jewish memorials and monuments with me when we travel. But more than that, you're the strongest, most empathetic, most wonderful person and partner I could have hoped to be with. At our wedding, you recited (in Hebrew!), "I am my beloved and my beloved is mine." I'm so lucky to be yours. I love you.

NOTES

Introduction

1. Rebecca King, phone interview, August 12, 2021.
2. "Jerusalem is Israel's Capital, says Donald Trump," BBC News, December 6, 2017, https://www.bbc.com/news/world-us-canada-42259443.
3. Allison Kaplan Sommer, "Trump Calls Biden a 'Servant of the Globalists,' Using Term Viewed as Antisemitic Dog Whistle," Haaretz, October 21, 2020, https://www.haaretz.com/us-news/.premium-jewish-democrats-attack-trump-for-calling-biden-a-servant-of-the-globalists-1.9250569.
4. https://twitter.com/adl/status/972235098853724161?lang=en.
5. See, for example, Josefin Dolstein, "Miriam Adelson Hopes There Will Be a Biblical 'Book of Trump,'" Times of Israel, June 28, 2019, https://www.timesofisrael.com/miriam-adelson-hopes-there-will-be-a-biblical-book-of-trump/.
6. Eric H. Yoffe, "The Republican Jewish Coalition's Shocking, Shameful Appeasement of Donald Trump," Haaretz, October 9, 2020, https://www.haaretz.com/us-news/the-republican-jewish-coalition-s-shocking-shameful-appeasement-of-donald-trump-1.9215920?lts=1636141065257.
7. See, for example, Brian Stewart and Quinn Scanlan, "Perdue's Campaign Deletes Ad That Enlarges Jewish Opponent's Nose, Insists It Was Accident," ABC News, ABC, July 28, 2020, https://abcnews.go.com/Politics/perdues-campaign-deletes-ad-enlarges-jewish-opponents-nose/story?id=72039950.
8. See, for example, Alex Kane, "How the ADL's Israel Advocacy Undermines Its Civil Rights Work," Jewish Currents, February 8, 2021, https://jewishcurrents.org/how-the-adls-israel-advocacy-undermines-its-civil-rights-work/.
9. See, for example, David H. Schanzer, "Most Jews Don't Vote for Trump

Because We Don't Share His Values," *Hill*, August 22, 2019, https://thehill .com/opinion/civil-rights/458428-most-jews-dont-vote-for-trump-because -we-dont-share-his-values.

10. See, for example, "How Not to Fight Antisemitism," *Jewish Currents*, April 5, 2021, https://jewishcurrents.org/how-not-to-fight-antisemitism.

11. "Biden, Buttigieg and Booker Confronted by Jewish Anti-occupation Activists," *Haaretz*, July 15, 2019, https://www.haaretz.com/us-news/biden -buttigieg-tell-ifnotnow-activists-they-support-an-end-to-israeli-occupation-1 .7501062.

12. https://twitter.com/benshapiro/status/133918830073352192.

13. "How the Faithful Voted," Pew Research Center, November 10, 2008, https:// www.pewforum.org/2008/11/05/how-the-faithful-voted/.

14. Ben Shapiro, "Jews in Name Only," CNSNews, Media Research Center, May 26, 2011, https://www.cnsnews.com/blog/ben-shapiro/jews-name-only.

15. Emily Tamkin, "Criticizing Mike Bloomberg in an Age of Antisemitism," *New Republic*, February 18, 2020, https://newrepublic.com/article/156592 /criticizing-michael-bloomberg-age-anti-semitism.

16. Steven Davidson, "Progressive Jewish Millennials Are Returning to Their Roots—Thanks to Trump," *Times of Israel*, March 16, 2018, https://www .timesofisrael.com/progressive-jewish-millennials-are-returning-to-their -roots-thanks-to-trump/.

17. Anshel Pfeffer, "Racism, Hate and Violence Are Jewish Values, Too," *Haaretz*, May 14, 2021, https://www.haaretz.com/israel-news/.premium .HIGHLIGHT-racism-hate-and-violence-are-jewish-values-too-1.9810151.

18. "How Stephen Miller Became the Architect of Trump's Immigration Policies," *Fresh Air*, NPR, August 24, 2020, https://www.npr.org/2020 /08/24/905457673/how-stephen-miller-became-the-architect-of-trumps -immigration-policies.

19. David Bernstein, "How Critical Social Justice Ideology Fuels Antisemitism," eJewish Philanthropy, September 3, 2021, https://ejewishphilanthropy.com /critical-social-justice-antisemitism/.

20. "Anti-Semitism," *Britannica*, https://www.britannica.com/topic/anti-Semi tism.

21. Gervase Phillips, "Antisemitism: How the Origins of History's Oldest Hatred Still Hold Sway Today," the *Conversation*, February 27, 2018, https://the conversation.com/antisemitism-how-the-origins-of-historys-oldest-hatred -still-hold-sway-today-87878.

22. Hasia Diner, *A New Promised Land: A History of Jews in America* (New York: Oxford University Press, 2000, 2003), 89.

23. Daniel Tamkin, phone interview, August 8, 2020.

24. Emily Tamkin, "Antisemitism in the Time of Trump," *New Statesman*, October 30, 2020, https://www.newstatesman.com/us-election-2020/2020/10 /anti-semitism-time-trump.

25. "Antisemitic Incidents Hit All-Time High in 2019," Anti-Defamation League, May 12, 2020, https://www.adl.org/news/press-releases/antisemitic-inci dents-hit-all-time-high-in-2019.

26. *The State of Antisemitism in America 2020: AJC's Survey of American Jews*, Amer-

ican Jewish Committee, https://www.ajc.org/AntisemitismReport2020/Survey-of-American-Jews.

27. See, for example, Karen Lehrman Bloch, "Left-Wing Antisemitism Is Insidious, Therefore Dangerous," *Jewish Journal*, April 16, 2020, https://jewish journal.com/commentary/columnist/314266/the-virus-anti-semitism/.

28. Emily Tamkin, "How Should US Antisemitism Be Defined in the Biden Era," *New Statesman*, February 4, 2021, https://www.newstatesman.com/world /americas/north-america/2021/02/how-should-us-anti-semitism-be-defined -biden-era-0.

29. Ofer Aderet, "Renowned Jewish Historian: Stop Using the Term 'Antisemitism,'" *Haaretz*, September 29, 2020, https://www.haaretz.com/israel -news/.premium-renowned-jewish-historian-stop-using-the-term-anti semitism-1.9193263.

30. Ron Kampeas, "Merrick Garland in AG Hearing: 'Senator, I'm a Pretty Good Judge of What an Antisemite Is,'" *Jewish Telegraphic Agency*, February 22, 2021, https://www.jta.org/quick-reads/merrick-garland-tears-up-describing -why-he-wants-to-serve-as-attorney-general-the-protection-us-gave-his-family.

31. Josh Perelman, phone interview, September 13, 2020.

32. Ruth Boehl, phone interview, December 10, 2020.

Chapter 1: Foreign Jews

1. Tamkin family recording.

2. Joellyn Zollman, "Jewish Immigration to America: Three Waves," My Jewish Learning, https://www.myjewishlearning.com/article/jewish-immigration -to-america-three-waves/.

3. Andrew Glass, "U.S. Enacts First Immigration Law, March 26, 1790," *Politico*, March 26, 2012, https://www.politico.com/story/2012/03/the-united -states-enacts-first-immigration-law-074438.

4. Michael Feldberg, "Judah Benjamin," Jewish Virtual Library, https://www .jewishvirtuallibrary.org/judah-benjamin.

5. Adam Serwer, *The Cruelty Is the Point* (New York: One World, Penguin Random House, 2021), 162.

6. Hasia Diner, *A New Promised Land* (New York: Oxford University Press, 2000, 2003), 38.

7. Diner, *New Promised Land*, 43.

8. Beverly Engle, phone interview, August 10, 2020.

9. "From Haven to Home: The Kishinev Massacre," Jewish Virtual Library, https://www.jewishvirtuallibrary.org/the-kishinev-massacre-judaic-treas ures.

10. Diner, *New Promised Land*, 44.

11. Diner, *New Promised Land*, 44.

12. Diner, *New Promised Land*, 52–53.

13. Peter Bacon, phone interview, September 13, 2020.

14. Sarah Colt and Thomas Jennings, "God in America: 'A New Light,'" PBS, 2010, https://www.pbs.org/godinamerica/transcripts/hour-four.html.

15. Max Winter, phone interview, March 8, 2021.

16. Diner, *New Promised Land*, 50, and Andrew Silow-Carroll, "An Old World Rabbi's Sad Encounter with New York Jewry," *New York Jewish Week*, December 15, 2020, https://jewishweek.timesofisrael.com/an-old-world-rabbis-sad-encounter-with-new-york-jewry/.

17. Diner, *New Promised Land*, 47.

18. Jonathan Sarna, *American Judaism* (New Haven, CT: Yale University Press, 2004, 2019), 192.

19. Diner, *New Promised Land*, 48–49.

20. "Declaration of Principles, 'The Pittsburgh Platform'—1885," Central Conference of American Rabbis, https://www.ccarnet.org/rabbinic-voice/platforms/article-declaration-principles/.

21. Eric Goldstein, *The Price of Whiteness: Jews, Race, and American Identity* (Princeton, NJ: Princeton University Press, 2006), 89.

22. Goldstein, *The Price of Whiteness*, 90.

23. Goldstein, *The Price of Whiteness*, 94–95.

24. Zev Chafets, "The Sy Empire," *New York Times Magazine*, October 14, 2007, https://www.nytimes.com/2007/10/14/magazine/14syrians-t.html.

25. Aviva Ben-Ur, *Sephardic Jews in America: A Diasporic History* (New York: New York University Press, 2009), 115.

26. Ben-Ur, *Sephardic Jews in America*, 1.

27. Ben-Ur, *Sephardic Jews in America*, 108.

28. Devin Naar, phone interview, October 1, 2020.

29. Ben-Ur, *Sephardic Jews in America*, 33.

30. Diner, *New Promised Land*, 65.

31. Sarna, *American Judaism*, 251.

32. Goldstein, *The Price of Whiteness*, 102–3.

33. Goldstein, *The Price of Whiteness*, 104–7.

34. Goldstein, *The Price of Whiteness*, 98.

35. Chafets, "The Sy Empire."

36. Goldstein, *The Price of Whiteness*, 99.

37. Keren McGinity, *Still Jewish: A History of Women and Intermarriage in America* (New York: New York University Press, 2009), 24.

38. Goldstein, *The Price of Whiteness*, 100–102.

39. McGinity, *Still Jewish*, 27.

40. Norman H. Finkelstein, *American Jewish History: A JPS Guide* (New York: Jewish Publication Society, 2007), 120–21.

41. Sarna, *American Judaism*, 214.

42. Sarna, *American Judaism*, 215.

43. Sarna, *American Judaism*, 224–25.

44. Sarna, *American Judaism*, 228.

45. Sarna, *American Judaism*, 226.

46. Sarna, *American Judaism*, 223.

47. Daniel Blatman, "Bund," *YIVO Encyclopedia of Jews in Eastern Europe*, https://yivoencyclopedia.org/article.aspx/bund.

48. Diner, *New Promised Land*, 57.

49. "About Us," *Forward*, https://forward.com/about-us/.

50. Ann Toback, phone interview, August 26, 2020.

51. Jia Lynn Yang, *One Mighty and Irresistible Tide: The Epic Struggle over American Immigration, 1924–1965* (New York: W. W. Norton, 2020), 36.

52. Hasia Diner, *Jews of the United States* (Berkeley: University of California Press, 2004), 164–65.

53. "Maryland's 'Jew Bill,'" Jewish Museum of Maryland, May 15, 2020, https://jewishmuseummd.org/marylands-jew-bill/.

54. Diner, *Jews of the United States*, 162–63.

55. Yang, *One Mighty and Irresistible Tide*, 23.

56. Goldstein, *The Price of Whiteness*, 129.

57. Sarna, *American Judaism*, 221.

58. Sarna, *American Judaism*, 223.

59. Yang, *One Mighty and Irresistible Tide*, 24–25.

60. Goldstein, *The Price of Whiteness*, 126.

61. Goldstein, *The Price of Whiteness*, 124.

62. Yang, *One Mighty and Irresistible Tide*, 26.

63. Yang, *One Mighty and Irresistible Tide*, 28–29.

64. Yang, *One Mighty and Irresistible Tide*, 43.

65. Yang, *One Mighty and Irresistible Tide*, 10.

66. Yang, *One Mighty and Irresistible Tide*, 54.

67. Yang, *One Mighty and Irresistible Tide*, 42.

68. 1920 United States Federal Census, Massachusetts, Suffolk, Boston Ward 15, District 0396.

69. Yang, *One Mighty and Irresistible Tide*, 42–43.

70. Yang, *One Mighty and Irresistible Tide*, 58.

71. Nurith Aizenman, "Trump Wishes We Had More Immigrants from Norway. Turns Out We Once Did," NPR, January 12, 2018, https://www.npr.org/sections/goatsandsoda/2018/01/12/577673191/trump-wishes-we-had-more-immigrants-from-norway-turns-out-we-once-did.

72. Yang, *One Mighty and Irresistible Tide*, 44–45.

73. Yang, *One Mighty and Irresistible Tide*, 58.

74. Finkelstein, *American Jewish History: A JPS Guide*, 130.

75. Tamkin family recording.

76. Daniel Tamkin, phone interview, August 8, 2020.

77. Engle, phone interview.

78. Finkelstein, *American Jewish History: A JPS Guide*, 130.

79. Diner, *New Promised Land*, 91–92.

80. Diner, *New Promised Land*, 75–76.

81. Diner, *New Promised Land*, 82–84.

82. See, for example, https://mobile.twitter.com/rabbisandra/status/1372855852420456451?lang=bg, and Anthony Russell, "'Go Down, Moses': Engaging with My Complex Musical Heritage at Passover," Tablet, April 7, 2014, https://www.tabletmag.com/sections/community/articles/passover-negro-spirituals.

83. Sarna, *American Judaism*, 255–57.

84. Sarna, *American Judaism*.

85. Goldstein, *The Price of Whiteness*, 128.

86. Goldstein, *The Price of Whiteness*, 164.

87. Sarna, *American Judaism*, 265–66.

88. Goldstein, *The Price of Whiteness*, 124.

89. Gary Phillip Zola and Marc Dollinger, eds., *American Jewish History* (Waltham, MA: Brandeis University Press, 2014), 228–30.

90. Yang, *One Mighty and Irresistible Tide*, 67.

91. Leonard Dinnerstein, "Jews and the New Deal," *American Jewish History* 72, no. 4 (June 1983): 461–76.

92. Goldstein, *The Price of Whiteness*, 167.

93. Goldstein, *The Price of Whiteness*, 171.

94. Goldstein, *The Price of Whiteness*, 174.

95. Goldstein, *The Price of Whiteness*, 184.

96. Dinnerstein, "Jews and the New Deal."

97. Goldstein, *The Price of Whiteness*, 191.

98. Arlette Sanders, phone interview, September 22, 2020.

99. Mae Ngai, *Impossible Subjects: Illegal Aliens and the Making of Modern America* (Princeton, NJ: Princeton University Press, 2004), 235.

100. Sarna, *American Judaism*, 265.

101. Finkelstein, *American Jewish History: A JPS Guide*, 133–34.

102. Finkelstein, *American Jewish History: A JPS Guide*, 132.

103. Finkelstein, *American Jewish History: A JPS Guide*, 136.

104. Yang, *One Mighty and Irresistible Tide*, 74–75.

105. Goldstein, *The Price of Whiteness*, 192.

106. Laurent Bouzereau, *Five Came Back* (2017, Netflix).

107. Deborah Dash Moore, *GI Jews: How World War II Changed a Generation* (Cambridge, MA: Belknap Press of Harvard University Press, 2004), 25–26.

108. Moore, *GI Jews*, 32.

109. Moore, *GI Jews*, 33–36.

110. Moore, *GI Jews*, 66–67.

111. Moore, *GI Jews*, 98.

112. Moore, *GI Jews*, 113–14.

113. Yang, *One Mighty and Irresistible Tide*, 74.

114. https://www.nytimes.com/1942/11/25/archives/himmler-program-kills-polish-jews-slaughter-of-250000-in-plan-to.html.

115. Yang, *One Mighty and Irresistible Tide*, 87–88.

116. Diner, *New Promised Land*, 41–42.

117. Yang, *One Mighty and Irresistible Tide*, 84–85.

118. Zola and Dollinger, *American Jewish History*, 256–57.

119. Sarna, *American Judaism*, 262–63.

120. Diner, *New Promised Land*, 89.

121. Bouzereau, *Five Came Back*.

122. Kenneth Jacobson, phone interview, September 13, 2020.

123. Sarna, *American Judaism*, 271.

124. Sarna, *American Judaism*, 271.

125. Sarna, *American Judaism*, 268.

126. Sarna, *American Judaism*, 267.

127. Sarna, *American Judaism*, 269–71.

128. Tamkin, phone interview.

Chapter 2: White and Red Jews

1. Karen Brodkin, *How Jews Became White Folks and What That Says About Race in America* (New Brunswick, NJ: Rutgers University Press, 1998, 2010), 142–43.

2. https://barbie.mattel.com/shop/en-us/ba/all-signature-dolls/barbie-dream-house-1962-repro-with-doll-gnc38.

3. Aviva Ben-Ur, *Jewish Autonomy in a Slave Society: Suriname in the Atlantic World, 1651–1825* (Philadelphia: University of Pennsylvania Press, 2020), 163.

4. "Lewis Charles Levin," Jewish Virtual Library, https://www.jewishvirtuallibrary.org/lewis-charles-levin.

5. Norman H. Finkelstein, "Jews in the Civil War," My Jewish Learning, https://www.myjewishlearning.com/article/jews-in-the-civil-war/.

6. Brodkin, *How Jews Became White Folks*, 76.

7. See, for example, Eric Goldstein, *The Price of Whiteness: Jews, Race, and American Identity* (Princeton, NJ: Princeton University Press, 2006), 122.

8. Goldstein, *The Price of Whiteness*, 95–96.

9. Arthur Liebman, *Jews and the Left* (New York: John Wiley and Sons, 1979), 34–36.

10. Goldstein, *The Price of Whiteness*, 161.

11. Goldstein, *The Price of Whiteness*, 110–15.

12. Goldstein, *The Price of Whiteness*, 161.

13. Goldstein, *The Price of Whiteness*, 110.

14. "NAACP: A Century in the Fight for Freedom," https://www.loc.gov/exhibits/naacp/founding-and-early-years.html.

15. Goldstein, *The Price of Whiteness*, 147.

16. Goldstein, *The Price of Whiteness*, 153–54.

17. Goldstein, *The Price of Whiteness*, 102–3.

18. Goldstein, *The Price of Whiteness*, 164.

19. Deborah Dash Moore, *GI Jews: How World War II Changed a Generation* (Cambridge, MA: Belknap Press of Harvard University Press, 2004), 261.

20. Robert Levinson, "Many Black World War II Veterans Were Denied Their GI Bill Benefits. Time to Fix That.," *War on the Rocks*, September 11, 2020, https://warontherocks.com/2020/09/many-black-world-war-ii-veterans-were-denied-their-gi-bill-benefits-time-to-fix-that/.

21. Rachel Kranson, *Ambivalent Embrace: Jewish Upward Mobility in Postwar America* (Chapel Hill: University of North Carolina Press, 2017), 47.

22. Hasia Diner, *A New Promised Land: A History of Jews in America* (New York: Oxford University Press, 2000, 2003), 94.

23. Kranson, *Ambivalent Embrace*, 56.

24. Daniel Tamkin, phone interview, August 8, 2020.

25. "President Eisenhower Signs 'In God We Trust' into Law," History.com, https://www.history.com/this-day-in-history/president-eisenhower-signs-in-god-we-trust-into-law.

26. Diner, *New Promised Land*, 99.

27. Kranson, *Ambivalent Embrace*, 71.

28. Kranson, *Ambivalent Embrace*, 71.

29. Hasia Diner, *Jews of the United States* (Berkeley: University of California Press, 2004), 293–98.

30. Jonathan Sarna, *American Judaism* (New Haven, CT: Yale University Press, 2004, 2019), 287.

31. Kranson, *Ambivalent Embrace*, 89–90.

32. Kranson, *Ambivalent Embrace*, 73.

33. Kranson, *Ambivalent Embrace*, 75–76.

34. Diner, *New Promised Land*, 104–5.

35. Diner, *New Promised Land*, 96.

36. Diner, *New Promised Land*, 104–5.

37. Diner, *Jews of the United States*, 261–65.

38. Kranson, *Ambivalent Embrace*, 44–45.

39. Kranson, *Ambivalent Embrace*, 61.

40. Allison Perlman, "Hollywood blacklist," *Britannica*, https://www.britannica.com/topic/Hollywood-blacklist.

41. Hasia Diner, Zoom interview, February 23, 2021.

42. Robert Meeropol, phone interview, February 12, 2021.

43. Emily Tamkin, "The Executed Innocent: Why Justice for Ethel Rosenberg Still Matters," *New Statesman*, March 26, 2021, https://www.newstatesman.com/world/americas/north-america/2021/03/executed-innocent-why-justice-ethel-rosenberg-still-matters.

44. "The Rosenberg Trial," Atomic Heritage Foundation, April 25, 2018, http://www.atomicheritage.org/history/rosenberg-trial.

45. Robert D. McFadden, "David Greenglass, the Brother Who Doomed Ethel Rosenberg, Dies at 92," *New York Times*, October 14, 2014, https://www.nytimes.com/2014/10/15/us/david-greenglass-spy-who-helped-seal-the-rosenbergs-doom-dies-at-92.html.

46. Sam Roberts, "Figure in Rosenberg Case Admits to Soviet Spying," *New York Times*, September 11, 2008, https://www.nytimes.com/2008/09/12/nyregion/12spy.html.

47. Meeropol, phone interview.

48. Gabrielle Bruney, "Roy Cohn Was an Infamous Political Fixer Who Made President Trump 'From Beyond the Grave,'" *Esquire*, June 18, 2020, https://www.esquire.com/entertainment/a29177110/wheres-my-roy-cohn-matt-tyrnauer-donald-trump-interview/.

49. Diner, *New Promised Land*, 102.

50. "Jim Hoberman's Oral History," Yiddish Book Center, June 26, 2011, https://www.yiddishbookcenter.org/collections/oral-histories/interviews/woh-fi-0000130/jim-hoberman-2011.

51. John, phone interview, September 8, 2021.

52. Kranson, *Ambivalent Embrace*, 17–27.

53. Joseph Dorman, *Sholem Aleichem: Laughing in the Darkness* (New York: Filmakers Library, 2012).

54. Kranson, *Ambivalent Embrace*, 21–23.

55. Kranson, *Ambivalent Embrace*, 30.

56. Aviva Ben-Ur, *Sephardic Jews in America: A Diasporic History* (New York: New York University Press, 2009), 35.

57. Brodkin, *How Jews Became White Folks*, 145–51.
58. Kranson, *Ambivalent Embrace*, 47–51.
59. "US Presidential Elections: Jewish Voting Record," Jewish Virtual Library, https://www.jewishvirtuallibrary.org/jewish-voting-record-in-u-s-presiden tial-elections.
60. Kranson, *Ambivalent Embrace*, 47–51.
61. Sarna, *American Judaism*, 314.
62. Kranson, *Ambivalent Embrace*, 116–17.
63. Tamkin, phone interview.
64. Kranson, *Ambivalent Embrace*, 114.
65. Kranson, *Ambivalent Embrace*, 118.
66. Tamkin, phone interview.
67. Keren McGinity, *Still Jewish: A History of Women and Intermarriage in America* (New York: New York University Press, 2009), 4.
68. Dorman, *Sholem Aleichem: Laughing in the Darkness*.
69. Samira K. Mehta, *Beyond Chrismukkah: The Christian-Jewish Interfaith Family in the United States* (Chapel Hill: University of North Carolina Press), 24.
70. Kranson, *Ambivalent Embrace*, 118–119.
71. Brodkin, *How Jews Became White Folks*, 160–63.
72. Brodkin, *How Jews Became White Folks*, 166–67.
73. Kranson, *Ambivalent Embrace*, 100–104.
74. Brodkin, *How Jews Became White Folks*, 131–32.
75. Kranson, *Ambivalent Embrace*, 112.
76. Max, phone interview, August 5, 2021.

Chapter 3: Zionist Jews

1. Hasia Diner, *A New Promised Land: A History of Jews in America* (New York: Oxford University Press, 2000, 2003), 62.
2. "Jewish & Non-Jewish Population of Israel/Palestine," Jewish Virtual Library, https://www.jewishvirtuallibrary.org/jewish-and-non-jewish-population-of -israel-palestine-1517-present.
3. Diner, *New Promised Land*, 62.
4. Diner, *New Promised Land*, 88.
5. Norman H. Finkelstein, *American Jewish History: A JPS Guide* (New York: Jewish Publication Society, 2007), 121.
6. Julian Cardillo, "Christmas Trees in Jewish Homes: A Brief History," *Brandeis NOW*, December 16, 2019, https://www.brandeis.edu/now/2019/december /christmas-trees-jewish-homes.html.
7. Finkelstein, *American Jewish History: A JPS Guide*, 121.
8. Jonathan Sarna, *American Judaism* (New Haven, CT: Yale University Press, 2004, 2019), 250–51.
9. Sarna, *American Judaism*, 253–54.
10. Diner, *New Promised Land*, 88.
11. Finkelstein, *American Jewish History: A JPS Guide*, 124.
12. Shaul Magid, "Liberal Zionism Is Dying. Will Foregoing the Jewish State

Save It?," *+972 Magazine*, August 10, 2021, https://www.972mag.com/haifa -republic-book-review-liberal-zionism/.

13. Calvin Goldscheider and Alan S. Zuckerman, "The Formation of Jewish Political Movements in Europe," *Modern Judaism* 4, no. 1 (February 1984): 83–104. Published by Oxford University Press.

14. Kalman Weiser, "Folkists," *YIVO Encyclopedia of Jews in Eastern Europe*, https://yivoencyclopedia.org/article.aspx/Folkists.

15. See "Origins and Evolution of the Palestine Problem: 1917–1947 (Part I)," United Nations, https://www.un.org/unispal/history2/origins-and -evolution-of-the-palestine-problem/part-i-1917–1947/, and "British Palestine Mandate," British White Papers, https://www.jewishvirtuallibrary.org /the-british-white-papers.

16. Daniel Tamkin, phone interview, August 8, 2020.

17. Finkelstein, *American Jewish History: A JPS Guide*, 145.

18. Anshel Pfeffer, "End U.S. 'Aid' to Israel. For the Sake of American Jews," *Haaretz*, May 21, 2021, https://www.haaretz.com/us-news/.premium.HI GHLIGHT-end-u-s-aid-to-israel-for-the-sake-of-american-jews-1.9829349.

19. Peter Beinart, *The Crisis of Zionism* (New York: Henry Holt, 2012), 13.

20. Beinart, *The Crisis of Zionism*.

21. Hussein Ibish, "A 'Catastrophe' That Defines Palestinian Identity," *Atlantic*, May 14, 2018, https://www.theatlantic.com/international/archive/2018 /05/the-meaning-of-nakba-israel-palestine-1948-gaza/560294/.

22. Sheryl Gay Stolberg, "A 'Calming Feeling,' a Frenzy and a New Front in the War over Anti-Semitism," *New York Times*, May 13, 2019, https://www.ny times.com/2019/05/13/us/politics/rashida-tlaib-holocaust.html.

23. "Rep. Rashida Tlaib on Growing Up in Detroit, Holocaust Comments and Fighting Poverty," *Late Night with Seth Meyers*, NBC, May 14, 2019, https:// www.youtube.com/watch?v=gDRHn3R2kc4&t=318s.

24. Arlette Sanders, phone interview, September 22, 2020.

25. Sarna, *American Judaism*, 337.

26. Sarna, *American Judaism*, 335–38.

27. Diner, *New Promised Land*, 103.

28. Daniel Gordis, *We Stand Divided: The Rift Between American Jews and Israel* (New York: ECCO, an imprint of HarperCollins, 2019), 40–43.

29. Gordis, *We Stand Divided*, 67.

30. Sarna, *American Judaism*, 296.

31. Hasia Diner, *Jews of the United States* (Berkeley: University of California Press, 2004), 295–96.

32. Sanders, phone interview.

33. Michael E. Staub, *Torn at the Roots: The Crisis of Jewish Liberalism in Postwar America* (New York: Columbia University Press, 2002), 63.

34. Sarna, *American Judaism*, 335–36.

35. Diner, *Jews of the United States*, 322–23.

36. Diner, *Jews of the United States*, 323.

37. Max Lewkowicz, *Fiddler: A Miracle of Miracles* (2019, Roadside Attractions, Samuel Goldwyn Films).

38. "Warsaw Ghetto Uprising," *Holocaust Encyclopedia*, United States Holocaust

Memorial Museum, https://encyclopedia.ushmm.org/content/en/article/warsaw-ghetto-uprising.

39. Diner, *Jews of the United States*, 324–25.
40. Sarna, *American Judaism*, 335–38.
41. Tamkin, phone interview.
42. Gordis, *We Stand Divided*, 14–15.
43. Diner, *Jews of the United States*, 322–24.
44. Diner, *Jews of the United States*, 324.
45. Tamkin, phone interview.
46. Abraham Riesman, "My Grandfather the Zionist," *New York Magazine*, June 23, 2021, https://nymag.com/intelligencer/2021/06/my-grandfather-the-zionist.html.
47. Tamkin, phone interview.
48. Jo-Ann Mort, Zoom interview, December 17, 2020.
49. Riesman, "My Grandfather the Zionist."

Chapter 4: Civil Rights Jews

1. "Pray with Your Feet," Central Synagogue, New York City, January 12, 2014, https://www.centralsynagogue.org/news/pray-with-your-feet. Heschel is sometimes quoted as having said that he felt his legs were praying.
2. "Negro Marchers from Selma Wear 'Yarmulkes' in Deference to Rabbis," *Jewish Telegraphic Agency*, March 23, 1965, https://www.jta.org/archive/negro-marchers-from-selma-wear-yarmulkes-in-deference-to-rabbis.
3. Shira Feder, "Georgia Senate Winner Raphael Warnock: Heschel and King 'Are Smiling in This Moment,'" *Jewish Telegraphic Agency*, January 6, 2021, https://www.sun-sentinel.com/florida-jewish-journal/fl-jj-georgia-senate-winner-raphael-warnock-heschel-king-smiling-20210106-flahl4oy3veg3o3ubigeggq2ym-story.html.
4. Rachel Kranson, *Ambivalent Embrace: Jewish Upward Mobility in Postwar America* (Chapel Hill: University of North Carolina Press, 2017), 53.
5. Marc Dollinger, *Black Power, Jewish Politics: Reinventing the Alliance in the 1960s* (Waltham, MA: Brandeis University Press, 2018), 28.
6. Hasia Diner, *Jews of the United States* (Berkeley: University of California Press, 2004), 268–69.
7. "Murder in Mississippi," *American Experience*, PBS, https://www.pbs.org/wgbh/americanexperience/features/freedomsummer-murder/.
8. Wayne King, "FILM; Fact v Fiction in Mississippi," *New York Times*, December 4, 1988, https://www.nytimes.com/1988/12/04/movies/film-fact-vs-fiction-in-mississippi.html.
9. Michael E. Staub, *Torn at the Roots: The Crisis of Jewish Liberalism in Postwar America* (New York: Columbia University Press, 2002), 87.
10. Staub, *Torn at the Roots*, 87–88.
11. Kranson, *Ambivalent Embrace*, 53–54.
12. Susannah Heschel, phone interview, March 17, 2021.
13. Kranson, *Ambivalent Embrace*, 54.

14.   Staub, *Torn at the Roots*, 49.

15.   Liza Kaufman Hogan, phone interview, August 20, 2021.

16.   Valerie Frey, Kaye Kole, and Luciana Spracher, *Voices of Savannah: Selections from the Oral History Collection of the Savannah Jewish Archives* (Savannah, GA: Savannah Jewish Archives, 2004), 70.

17.   Staub, *Torn at the Roots*, 53–54.

18.   Dollinger, *Black Power, Jewish Politics*, 55.

19.   Dollinger, *Black Power, Jewish Politics*, 70–71.

20.   Staub, *Torn at the Roots*, 96.

21.   Staub, *Torn at the Roots*.

22.   Staub, *Torn at the Roots*, 99.

23.   Dollinger, *Black Power, Jewish Politics*, 62.

24.   Dollinger, *Black Power, Jewish Politics*, 69.

25.   Heschel, phone interview.

26.   Heschel, phone interview.

27.   Dollinger, *Black Power, Jewish Politics*, 34.

28.   Dollinger, *Black Power, Jewish Politics*, 32–33.

29.   Dollinger, *Black Power, Jewish Politics*, 85.

30.   James Baldwin, "Negroes Are Antisemitic Because They're Anti-White," *New York Times*, April 9, 1967, https://archive.nytimes.com/www.nytimes.com/books/98/03/29/specials/baldwin-antisem.html?_r=1.

31.   Judith Weisenfeld, "Race, Religion, and Black Jewish Identity in the Early Twentieth-Century U.S.," Jewish Studies-Fordham, October 12, 2021, https://www.youtube.com/watch?v=C5dLZZP0tUs.

32.   "Extremist Sects Within the Black Hebrew Israelite Movement," Anti-Defamation League, https://www.adl.org/resources/backgrounders/extremist-sects-within-the-black-hebrew-israelite-movement.

33.   Jia Lynn Yang, *One Mighty and Irresistible Tide: The Epic Struggle over American Immigration, 1924–1965* (New York: W. W. Norton, 2020), 176–83.

34.   Yang, *One Mighty and Irresistible Tide*, 243–260.

35.   Mae M. Ngai, *Impossible Subjects: Illegal Aliens and the Making of Modern America* (Princeton, NJ: Princeton University Press, 2004), 263–64.

36.   Keren McGinity, *Still Jewish: A History of Women and Intermarriage in America* (New York: New York University Press, 2009), 110.

37.   Sandra Lawson, phone interview, October 18, 2021.

38.   Anthony Russell, phone interview, October 26, 2021.

39.   Peter Beinart, *The Crisis of Zionism* (New York: Henry Holt, 2012), 14.

40.   "Friedman Said Seeking to Call West Bank 'Judea and Samaria' in Statements," *Times of Israel*, April 25, 2018, https://www.timesofisrael.com/friedman-seeking-to-call-west-bank-judea-and-samaria-in-official-statements/.

41.   Beinart, *The Crisis of Zionism*, 20.

42.   Diner, *Jews of the United States*, 335.

43.   "Stokely Carmichael," *Britannica*, https://www.britannica.com/biography/Stokely-Carmichael.

44.   Diner, *Jews of the United States*, 335.

45.   See, for example, Ofer Aderet, "'We Saw Jews with Hearts Like Germans': Moroccan Immigrants in Israel Warned Families Not to Follow," *Haaretz*,

July 10, 2021, https://www.haaretz.com/israel-news/.premium.HIGHLI GHT.MAGAZINE-horrified-by-racism-moroccan-immigrants-in-israel -warned-families-not-to-follow-1.9982152.

46. Diane Pien, "Israeli Black Panther Party (1971–1977)," BlackPast, July 2, 2018, https://www.blackpast.org/global-african-history/israeli-black-panther -party-1971–1977/.

47. "Ethiopian Jews: Israel's Second-Class Citizens?," *Deutsche Welle*, September 29, 2018, https://www.dw.com/en/ethiopian-jews-israels-second-class -citizens/a-45687623.

48. Name withheld, phone interview, March 3, 2021.

49. Diner, *Jews of the United States*, 335.

50. Dollinger, *Black Power, Jewish Politics*, 53.

51. Dollinger, *Black Power, Jewish Politics*, 94.

52. Daniel Gordis, *We Stand Divided: The Rift Between American Jews and Israel* (New York: ECCO, an imprint of HarperCollins, 2019), 69.

53. Staub, *Torn at the Roots*, 137.

54. Staub, *Torn at the Roots*, 142.

55. Staub, *Torn at the Roots*, 158–59.

56. Arthur Waskow, Zoom interview, January 8, 2021.

57. Staub, *Torn at the Roots*, 181–90.

58. Dollinger, *Black Power, Jewish Politics*, 105–6.

59. Dollinger, *Black Power, Jewish Politics*, 112–19.

60. Dollinger, *Black Power, Jewish Politics*, 125.

61. Dollinger, *Black Power, Jewish Politics*, 144–49.

62. Dollinger, *Black Power, Jewish Politics*, 116.

63. "Jimmy Carter and the Rise of Evangelical Voters," RetroReport, https:// www.retroreport.org/education/video/jimmy-carter-and-the-rise-of-evan gelical-voters/.

64. http://content.time.com/time/covers/0,16641,19771226,00.html.

65. Philip Bump, "Half of Evangelicals Support Israel Because They Believe It Is Important for Fulfilling End-Times Prophecy," *Washington Post*, May 14, 2018, https://www.washingtonpost.com/news/politics/wp/2018/05/14 /half-of-evangelicals-support-israel-because-they-believe-it-is-important-for -fulfilling-end-times-prophecy/.

66. Heschel, phone interview.

67. Bruce D. Haynes, *The Soul of Judaism: Jews of African Descent in America* (New York: New York University Press, 2018), 175.

68. Sandee Brawarsky, "Tracing Tevye's Cultural Footprint," *New York Jewish Week*, November 13, 2013, https://jewishweek.timesofisrael.com/tracing -tevyes-cultural-footprint/.

69. Haynes, *The Soul of Judaism*, 172.

70. Brawarsky, "Tracing Tevye's Cultural Footprint."

71. Max Lewkowicz, *Fiddler: A Miracle of Miracles* (2019, Roadside Attractions, Samuel Goldwyn Films).

72. McGinity, *Still Jewish*, 110–12.

73. McGinity, *Still Jewish*, 123.

74. Lila Corwin Berman, Kate Rosenblatt, and Ronit Stahl, "Continuity Crisis:

The History and Sexual Politics of an American Jewish Communal Project," *American Jewish History* 104, nos. 2/3 (April/July 2020): 173.

75. McGinity, *Still Jewish*, 165.
76. McGinity, *Still Jewish*, 126–27.
77. Staub, *Torn at the Roots*, 262.
78. Corwin Berman, Rosenblatt, and Stahl, "Continuity Crisis," 177.
79. Heschel, phone interview.
80. Hannah Dreyfus, "Steven M. Cohen, Shunned by Academy After Harassment Allegations, Makes Stealthy Comeback—and Provokes Uproar," *Forward*, March 23, 2021, https://forward.com/news/466437/steven-cohen-shunned -after-harassment-accusations-trying-to-make-a-stealth/.
81. Heschel, phone interview.

Chapter 5: Right-Wing Jews

1. Jacob Heilbrunn, *They Knew They Were Right: The Rise of the Neocons* (New York: Doubleday, 2008), 56.
2. Heilbrunn, *They Knew They Were Right*, 29–36.
3. Heilbrunn, *They Knew They Were Right*, 36–42.
4. See, for example, "The Rise of the Neocons," *Week*, January 8, 2015, https:// theweek.com/articles/528827/rise-neocons.
5. Heilbrunn, *They Knew They Were Right*, 81.
6. Norman Podhoretz, "My Negro Problem—and Ours," *Commentary*, February 1963, https://www.commentary.org/articles/norman-podhoretz/my-negro -problem-and-ours/.
7. Karen Brodkin, *How Jews Became White Folks and What That Says About Race in America* (New Brunswick, NJ: Rutgers University Press, 1998, 2010), 145–50.
8. Heilbrunn, *They Knew They Were Right*, 11–22.
9. Douglas Feith, personal interview, Washington, DC, 2019.
10. Norman Podhoretz, "Is It Good for the Jews?," *Commentary*, February 1972, https://www.commentary.org/articles/norman-podhoretz/is-it-good-for -the-jews/.
11. Benjamin Balint, *Running Commentary: The Contentious Magazine That Transformed the Jewish Left into the Neoconservative Right* (New York: PublicAffairs, 2010), 12–16.
12. Balint, *Running Commentary*, 20–21.
13. Balint, *Running Commentary*, 50–51.
14. Balint, *Running Commentary*, 57.
15. Balint, *Running Commentary*, 65–73.
16. Balint, *Running Commentary*, 72–73.
17. Balint, *Running Commentary*, 76.
18. Balint, *Running Commentary*, 84.
19. Balint, *Running Commentary*, 98–102.
20. Balint, *Running Commentary*, 106.
21. Balint, *Running Commentary*, 114.
22. Balint, *Running Commentary*, 118–21.

23. Balint, *Running Commentary*, 124–25.
24. Balint, *Running Commentary*, 126.
25. Balint, *Running Commentary*, 133.
26. Balint, *Running Commentary*, 131.
27. Lila Corwin Berman, *The American Jewish Philanthropic Complex: The History of a Multibillion-Dollar Institution* (Princeton, NJ: Princeton University Press, 2020), 160.
28. https://www.jewishdatabank.org/content/upload/bjdb/599/N-Jewish _American_Voting_Solomon_Project_2012_Main_Report.pdf.
29. Balint, *Running Commentary*, 211.
30. Heilbrunn, *They Knew They Were Right*, 161–64.
31. Feith, personal interview.
32. Balint, *Running Commentary*, 205.
33. Patrick Glynn, "Why an Arms Build-Up Is Morally Necessary," *Commentary*, February 1984, https://www.commentary.org/articles/patrick-glynn/why -an-american-arms-build-up-is-morally-necessary/.
34. Heilbrunn, *They Knew They Were Right*, 170–84.
35. Heilbrunn, *They Knew They Were Right*, 158.
36. Heilbrunn, *They Knew They Were Right*, 209.
37. Norman Podhoretz, "Neoconservatism: A Eulogy," American Enterprise Institute, January 15, 1996.
38. Balint, *Running Commentary*, 183.
39. Balint, *Running Commentary*, 185.
40. Michelle Mark, "Former CIA Officer Valerie Plame Wilson Apologizes for Tweeting a Story Blaming Jews for US Wars in the Middle East," Business Insider, September 21, 2017, https://www.businessinsider.com/valerie-plame -wilson-jews-tweet-wars-2017-9.
41. Corwin Berman, *The American Jewish Philanthropic Complex*, 157.
42. Balint, *Running Commentary*, 143.

Chapter 6: Laboring Jews

1. "Merchant of Venice," *Treasures in Full: Shakespeare in Quarto*, British Library, https://www.bl.uk/treasures/shakespeare/merchant.html.
2. Jonathan Sarna, *American Judaism* (New Haven, CT: Yale University Press, 2004, 2019), 217.
3. Daniel Tamkin, personal interview, August. 8, 2020.
4. Ed Delman, Zoom interview, October 7, 2020.
5. Elizabeth Tamkin, phone interview, August 13, 2021.
6. Josh Fitt, phone interview, September 16, 2020.
7. Name withheld, phone interview, August 4, 2021.
8. Devan Cole, "House Majority Leader Deletes Tweet Saying Soros, Bloomberg, Steyer Are Trying to 'Buy' Election," CNN, October 28, 2018, https:// www.cnn.com/2018/10/28/politics/tom-steyer-mccarthy-tweet/index .html.
9. "Fox News' Tucker Carlson Claims George Soros Is 'Remaking' America,"

Haaretz, June 13, 2019, https://www.haaretz.com/world-news/fox-news
-tucker-carlson-claims-george-soros-is-remaking-america-1.7362111.

10. Emily Tamkin, *The Influence of Soros: Politics, Power, and the Struggle for an Open Society* (New York: Harper, an imprint of HarperCollins, 2020), 2.

11. "Who Invented the Weekend?," BBC Bitesize, https://www.bbc.co.uk/bitesize/articles/zf22kmn.

12. "Shabbat as Social Reform," My Jewish Learning, https://www.myjewishlearning.com/article/shabbat-as-social-reform/.

13. Yuri Slezkine, *The Jewish Century* (Princeton, NJ: Princeton University Press, 2004), 213–15.

14. A. Gurwitz, "Perfectly Matched and Perfectly Timed," in *Atlantic Metropolis: An Economic History of New York* (Basingstoke: Palgrave Macmillan, 2019), 323–58.

15. Slezkine, *The Jewish Century*, 318.

16. Ann Toback, phone interview, August 26, 2020.

17. Hasia Diner, Zoom interview, February 23, 2021.

18. Michael E. Staub, *Torn at the Roots: The Crisis of Jewish Liberalism in Postwar America* (New York: Columbia University Press, 2002), 53.

19. "Lehman Brothers," *Encyclopedia of Alabama*, Alabama Humanities Alliance, May 21, 2009, http://encyclopediaofalabama.org/article/h-2160.

20. Jerry Muller, phone interview, May 27, 2021.

21. I. W. Burnham II, "Ethnic Bankers," *New York Times*, July 13, 1986, https://www.nytimes.com/1986/07/13/business/l-ethnic-bankers-525686.html.

22. "Michael Milken," *Britannica*, https://www.britannica.com/biography/Michael-R-Milken#ref260895.

23. Dan Mangan, "Trump Pardons Michael Milken, Face of 1980s Insider Trading Scandals," CNBC, February 18, 2020, https://www.cnbc.com/2020/02/18/trump-pardons-michael-milken-face-of-1980s-financial-scandals.html.

24. David Warsh, "Undercurrents of Religious, Cultural Antagonism in Drexel's Rise, Fall," *Washington Post*, February 21, 1990, https://www.washingtonpost.com/archive/business/1990/02/21/undercurrent-of-religious-cultural-antagonism-in-drexels-rise-fall/735af83d-5320–4985–925e-8a5cee03af93/.

25. David A. Kaplan, "Wall Street: A Greed Apart," *Newsweek*, October 13, 1991, https://www.newsweek.com/wall-street-greed-apart-204816.

26. "A History of the Rothschild Family," Investopedia, https://www.investopedia.com/updates/history-rothschild-family/.

27. Muller, phone interview.

28. https://www.sobpedro.com/our-history#:~:text=In%201949%2C%20Mr.,of%20the%20Border%20Drive%2DIn.

29. Rudy Maxa, "South of the Border Down Carolina Way," *Washington Post*, January 7, 1979, https://www.washingtonpost.com/archive/lifestyle/magazine/1979/01/07/south-of-the-border-down-carolina-way/3090a021-2fb5-457c-8fdb-b17ec4022d44/.

30. Sofia Lesnewski, "Filmmakers Explore Racial Impact of South of the Border in Paused Documentary," *Daily Tar Heel*, May 4, 2021, https://www.dailytarheel.com/article/2021/05/city-south-of-border.

31. Maxa, "South of the Border Down Carolina Way."

32. Lesnewski, "Filmmakers Explore Racial Impact of South of the Border in Paused Documentary."

33. Ann G. Wolfe, "The Invisible Jewish Poor," *Journal of Jewish Communal Service*, June 8, 1971, https://www.bjpa.org/content/upload/bjpa/the_/THE %20INVISIBLE%20JEWISH%20POOR.pdf.

34. Jonathan Hornstein, *Jewish Poverty in the United States: A Summary of Recent Research*, Harry and Jeanette Weinberg Foundation, February 2019, https://cdn.fedweb.org/fed-42/2892/jewish-poverty-in-the-united-states %2520Weinberg%2520Report.pdf.

35. Wolfe, "The Invisible Jewish Poor."

36. Cathy Free, "About a Third of Holocaust Survivors in the U.S. Live in Poverty. This Group Helps Them.," *Washington Post*, March 24, 2021, https:// www.washingtonpost.com/lifestyle/2021/03/24/holocaust-survivor -poverty-kavod/.

37. https://www.jewishfederations.org/.

38. Lila Corwin Berman, *The American Jewish Philanthropic Complex*, (Princeton, NJ: Princeton University Press, 2020), 133 and 161–163.

39. Lila Corwin Berman, phone interview, July 2, 2021.

40. Raphael Magarik, "Who Owns American Judaism?," *Jewish Currents*, June 1, 2021, https://jewishcurrents.org/who-owns-american-judaism.

41. Maayan Hoffman, "Federation and Its Future: Floundering or Flourishing?," eJewish Philanthropy, November 12, 2017, https://ejewishphilanthropy.com /jfna-and-its-future-floundering-or-flourishing/.

42. Corwin Berman, phone interview.

43. Keren McGinity, phone interview, July 4, 2021.

44. Kate Rosenblatt, phone interview, September 29, 2020.

45. Lila Corwin Berman, Kate Rosenblatt, and Ronit Stahl, "Continuity Crisis: The History and Sexual Politics of an American Jewish Communal Project," *American Jewish History* 104, nos. 2/3 (April/July 2020): 167–94.

46. Keren McGinity, *Still Jewish: A History of Women and Intermarriage in America* (New York: New York University Press, 2009), 6.

47. Helen Chernikoff, "The Pride of Matthew Bronfman's Philanthropy," eJewish Philanthropy, March 3, 2021, https://ejewishphilanthropy.com/matthew -bronfmans-philanthropy-is-about-pride/.

48. "Biography of Yitzhak Rabin," Rabin Center, http://www.rabincenter.org .il/Items/01097/Biography.pdf.

49. "Netanyahu Can't Wash His Hands of Incitement That Led to Rabin's Murder," *Haaretz*, November 13, 2016, https://www.haaretz.com/opinion/edit orial-netanyahu-can-t-wash-his-hands-of-incitement-1.5461189.

50. Zack Beauchamp, "What Were the Intifadas?," Vox, May 14, 2018, https:// www.vox.com/2018/11/20/18080066/israel-palestine-intifadas-first -second.

51. Michael Lipka, "7 Key Findings About Religion and Politics in Israel," Pew Research Center, March 8, 2016, https://www.pewresearch.org/fact-tank /2016/03/08/key-findings-religion-politics-israel/.

52. Laura E. Adkins and Ben Sales, "The Kids Are All Right-Wing: Why Israel's Younger Voters Are More Conservative," Jewish Telegraphic Agency,

April 11, 2019, https://www.timesofisrael.com/the-kids-are-all-right-wing-why-israels-younger-voters-are-more-conservative/.

53. Tom Mashberg, "Michael Steinhardt, Billionaire, Surrenders $70 Million in Stolen Relics," *New York Times*, December 6, 2021, https://www.nytimes.com/2021/12/06/arts/design/steinhardt-billionaire-stolen-antiquities.html.

54. "Adelson Gift of $25 Million a Year Provides Huge Boost to Birthright," Jewish Telegraphic Agency, February 8, 2007, https://www.jta.org/archive/adelson-gift-of-25-million-a-year-provides-huge-boost-to-birthright.

55. Erez Linn and ILH Staff, "To Celebrate Israel's 70th Anniversary, Adelsons Donate $70M to Taglit," *Israel Hayom*, April 17, 2018, https://www.israelhayom.com/2018/04/17/to-celebrate-israels-70th-anniversary-adelsons-donate-70m-to-taglit/.

56. Rachel Delia Benaim and Lazar Berman, "Sheldon Adelson Calls on US to Nuke Iranian Desert," *Times of Israel*, October 24, 2013, https://www.timesofisrael.com/sheldon-adelson-calls-on-us-to-nuke-iranian-desert/.

57. Barak Ravid, "Report: Adelson Offered Obama $1b to Develop Iron Dome for Israel," *Haaretz*, January 7, 2016, https://www.haaretz.com/israel-news/.premium-report-adelson-offered-obama-1b-to-develop-iron-dome-for-israel-1.5387565.

58. Amir Tibon, "This Powerful Adelson-Funded Israel Lobby Could Soon Rival AIPAC's Influence in Washington," *Haaretz*, October 31, 2017, https://www.haaretz.com/us-news/.premium-the-adelson-funded-group-that-could-rival-aipacs-influence-in-d-c-1.5461209.

59. Sheldon Adelson, "Sheldon Adelson: I Endorse Donald Trump for President," *Washington Post*, May 13, 2016, https://www.washingtonpost.com/opinions/sheldon-adelson-i-endorse-donald-trump-for-president/2016/05/12/ea89d7f0-17a0-11e6-aa55-670cabef46e0_story.html.

60. Robert D. McFadden, "Sheldon Adelson, Billionaire Donor to G.O.P. and Israel, Is Dead at 87," *New York Times*, January 12, 2021, https://www.nytimes.com/2021/01/12/business/sheldon-adelson-dead.html.

61. Amir Tibon, "Sheldon Adelson Donated $5 Million to Trump's Inauguration, Documents Reveal," April 19, 2017, https://www.haaretz.com/us-news/adelson-gave-5-million-to-trump-s-inauguration-team-1.5462556.

62. "Obituary: Sheldon Adelson, the Casino Magnate Who Moved an Embassy," *BBC News*, BBC, January 12, 2021, https://www.bbc.com/news/world-us-canada-47414850.

63. Josefin Dolstein, "Miriam Adelson Hopes There Will Be a Biblical 'Book of Trump,'" *Times of Israel*, June 28, 2019, https://www.timesofisrael.com/miriam-adelson-hopes-there-will-be-a-biblical-book-of-trump/.

64. Emily Tamkin, *The Influence of Soros*, 109.

65. Edward N. Luttwak, "My Meeting with George Soros," Tablet, June 18, 2020, https://www.tabletmag.com/sections/news/articles/meeting-with-george-soros.

66. Emily Tamkin, *The Influence of Soros*.

67. Luttwak, "My Meeting with George Soros."

68. Name withheld, phone interview, July 14, 2021.

69. Al Rosenberg, phone interview, August 23, 2021.
70. Name withheld, phone interview, August 18, 2021.
71. For example, Daniel Zemel, personal interview, Washington, DC, October 14, 2021.
72. https://twitter.com/tedcruz/status/1229919334539186177.
73. Emily Tamkin, "Criticizing Mike Bloomberg in an Age of Antisemitism," *New Republic*, February 18, 2020, https://newrepublic.com/article/156592/criticizing-michael-bloomberg-age-anti-semitism.
74. Michael Balsamo and Tom Hays, "Ponzi Schemer Bernie Madoff Dies in Prison at 82," Associated Press, April 14, 2021, https://abcnews.go.com/Business/wireStory/ap-source-ponzi-schemer-bernie-madoff-dies-prison-77065566.
75. Cindy Cardinal, phone interview, October 9, 2021.
76. Shelby Magid, phone interview, July 5, 2021.
77. Corwin Berman, phone interview.

Chapter 7: Refugee Jews

1. Olya, phone interview, February 11, 2021.
2. Ari L. Goldman, "Russian Jews Come to U.S. in Big Group," *New York Times*, September 29, 1989, https://www.nytimes.com/1989/09/29/nyregion/russian-jews-come-to-us-in-big-group.html.
3. Davar Ardalan, "For Persian Jews, America Means 'Religious Pluralism at Its Best,'" NPR, January 26, 2014, https://www.npr.org/sections/codeswitch/2014/01/26/260779898/for-persian-jews-america-means-religious-pluralism-at-its-best.
4. Yuri Slezkine, *The Jewish Century* (Princeton, NJ: Princeton University Press, 2004), 213–15.
5. "Joseph Stalin," Jewish Virtual Library, https://www.jewishvirtuallibrary.org/joseph-stalin.
6. "The Khrushchev Era," *Britannica*, https://www.britannica.com/place/Russia/The-Khrushchev-era-1953-64#ref422198.
7. Benjamin Nathans, "Alexander Volpin and the Origins of the Soviet Human Rights Movement," National Council for Eurasian and East European Research, 2007, https://www.ucis.pitt.edu/nceeer/2007_819–2g_Nathans.pdf.
8. Zinovy Zinik, "Alexander Ginzburg," *Guardian*, July 26, 2002, https://www.theguardian.com/news/2002/jul/26/guardianobituaries.booksobituaries.
9. Arina Ginzburg, personal interview, Paris, 2013.
10. See, for example, Gal Beckerman, *When They Come for Us, We'll Be Gone: The Epic Struggle to Save Soviet Jewry* (New York: First Mariner Books, 2011), 7.
11. Ludmilla Alexeyeva and Paul Goldberg, *The Thaw Generation: Coming of Age in the Post-Stalin Era* (Boston: Little, Brown, 1990), 280–81.
12. Benjamin Nathans, "Refuseniks and Rights Defenders: Jews and the Soviet Dissident Movement," in *From Europe's East to the Middle East: Israel's Russian and Polish Lineages*, ed. Kenneth B. Moss, Benjamin Nathans, and Taro Tsurumi (University of Pennsylvania Press: Philadelphia, 2021), 367.

13. Robert Zaretsky, "Realism No Excuse for Kissinger's Remark," RealClear World, January 9, 2011, https://www.realclearworld.com/articles/2011/01/09/realism_and_nixon_anti-semitism_99345.html.

14. Yaacov Ro'i, "Jackson-Vanik Amendment," *The YIVO Encyclopedia of Jews in Eastern Europe*, https://yivoencyclopedia.org/article.aspx/Jackson-Vanik_Amendment.

15. Beckerman, *When They Come for Us, We'll Be Gone*, 525–27.

16. Gold, "Soviet Jews in the United States."

17. https://www.hias.org/who/history.

18. Steven J. Gold, "Soviet Jews in the United States," *The American Jewish Yearbook* 94 (1994): 3–57, https://www.jstor.org/stable/i23605123.

19. *A Portrait of Jewish Americans*, Pew Research Center, October 1, 2013, https://www.pewforum.org/2013/10/01/jewish-american-beliefs-attitudes-culture-survey/.

20. Roman, phone interview, September 1, 2021.

21. Olga Khazan, "Why Soviet Refugees Aren't Buying Sanders's Socialism," *Atlantic*, April 12, 2016, https://www.theatlantic.com/politics/archive/2016/04/bernie-sanders-trump-russians/477045/.

22. Ethan Marcus, "Alejandro Mayorkas' Historic Nomination Is a Wake-Up Call: Stop Erasing Sepharadim," *Forward*, November 25, 2020, https://forward.com/opinion/459218/alejandro-mayorkass-historic-nomination-should-be-a-wake-up-call-for/.

23. Laura Limonic, *Kugel and Frijoles: Latin Jews in the United States* (Detroit: Wayne State University Press, 2019), 151.

24. See, for example, Beckerman, *When They Come for Us, We'll Be Gone*, 100–101.

25. Philip Reeves, "On Multiple Fronts, Russian Jews Reshape Israel," NPR, January 2, 2013, https://www.npr.org/2013/01/02/168457444/on-multiple-fronts-russian-jews-reshape-israel.

26. Rachel, phone interview, September 20, 2021.

27. Gennadyi Gurman, phone interview, September 6, 2021.

28. Olya, phone interview.

29. Davar Ardalan, "For Persian Jews, America Means 'Religious Pluralism at Its Best,'" https://www.npr.org/sections/codeswitch/2014/01/26/260779898/for-persian-jews-america-means-religious-pluralism-at-its-best.

30. Esther Amini, *Concealed: Memoir of a Jewish-Iranian Daughter Caught Between the Chador and America* (New York: Greenpoint Press, 2020).

31. Haideh Sahim, phone interview, August 6, 2021.

32. Julia, phone interview, July 23, 2021.

33. https://www.youtube.com/watch?v=NMjpppudNkk.

34. Julia, phone interview.

35. Ardalan, "For Persian Jews, America Means 'Religious Pluralism at Its Best.'"

36. Sarina Roffe, "Charismatic Rabbi Takes Reins of Manhattan Beach Synagogue," *Image*, June 2013, https://images.shulcloud.com/414/uploads/Article/rabbi-bitton.pdf.

37. Sahim, phone interview.

38. Mijal Bitton, "Many Jews of Color and Diverse Jews Are Politically Conservative—and Many Voted for Trump," Jewish Telegraphic Agency,

November 19, 2020, https://www.jta.org/2020/11/19/opinion/many-jews-of-color-and-diverse-jews-are-politically-conservative-and-many-voted-for-trump.

39. Anna Kaplan, phone interview, August 4, 2021.
40. Bitton, "Many Jews of Color and Diverse Jews Are Politically Conservative."
41. Kaplan, phone interview.
42. Philip Mehdipour, "Why Iranian Jews in America Love Trump," *Forward*, September 14, 2020, https://forward.com/opinion/454190/why-iranian-american-jews-love-trump/.
43. Maayan Malter, phone interview, August 5, 2021.
44. Julia, phone interview.
45. Jordan Kutzik, phone interview, September 17, 2020.
46. Sahim, phone interview.
47. Vladislav Davidzon, phone interview, September 2, 2021.

Chapter 8: "This Land Is Our Land" Jews

1. Alexia Underwood, "The Controversial US Jerusalem Embassy Opening, Explained," Vox, May 16, 2018, https://www.vox.com/2018/5/14/17340798/jerusalem-embassy-israel-palestinians-us-trump.
2. Quint Forgey, "'The Dawn of a New Middle East': Trump Celebrates Abraham Accords with White House Signing Ceremony," *Politico*, September 15, 2020, https://www.politico.com/news/2020/09/15/trump-abraham-accords-palestinians-peace-deal-415083.
3. Peter Baker and Julie Hirschfeld Davis, "U.S. Finalizes Deal to Give Israel $38 Billion in Military Aid," *New York Times*, September 13, 2016, https://www.nytimes.com/2016/09/14/world/middleeast/israel-benjamin-netanyahu-military-aid.html.
4. Krishnadev Calamur, "In Speech to Congress, Netanyahu Blasts 'A Very Bad Deal' with Iran," NPR, March 3, 2015, https://www.npr.org/sections/thetwo-way/2015/03/03/390250986/netanyahu-to-outline-iran-threats-in-much-anticipated-speech-to-congress.
5. "Full Text of US Envoy Samantha Power's Speech After Abstention on Anti-settlement Vote," *Times of Israel*, December 24, 2016, https://www.timesofisrael.com/full-text-of-us-envoy-samantha-powers-speech-after-abstention-on-anti-settlement-vote/.
6. "Here's Donald Trump's Presidential Announcement Speech," *Time*, June 16, 2015, https://time.com/3923128/donald-trump-announcement-speech/.
7. Nurith Aizenman, "Trump Wishes We Had More Immigrants from Norway. Turns Out We Once Did," NPR, January 12, 2018, https://www.npr.org/sections/goatsandsoda/2018/01/12/577673191/trump-wishes-we-had-more-immigrants-from-norway-turns-out-we-once-did.
8. Jodi Kantor, "For Kushner, Israel Policy May Be Shaped by the Personal," *New York Times*, February 11, 2017, https://www.nytimes.com/2017/02/11/us/politics/jared-kushner-israel.html.
9. Irene Katz Connelly and Talya Zax, "At Vice Presidential Debate, Pence

Invokes 'Jewish Grandchildren' in Defense of Trump," *Forward*, October 7, 2020, https://forward.com/news/456049/pence-jewish-grandchildren-trump -harris-vp-debate-white-nationalism/.

10.　Adam Serwer, *The Cruelty Is the Point: The Past, Present, and Future of Trump's America* (New York: One World, 2021), 146–47.

11.　David S. Glosser, "Stephen Miller Is an Immigration Hypocrite. I Know Because I'm His Uncle," *Politico*, August 13, 2018, https://www.politico.com /magazine/story/2018/08/13/stephen-miller-is-an-immigration-hypocrite -i-know-because-im-his-uncle-219351/.

12.　Liz Stark, "White House Policy Adviser Downplays Statue of Liberty's Famous Poem," CNN, August 3, 2017, https://www.cnn.com/2017/08/02 /politics/emma-lazarus-poem-statue-of-liberty/index.html.

13.　Ben Lorber, phone interview, August 24, 2021.

14.　"Jerusalem—The Upper City During the Second Temple Period," Israeli Ministry of Foreign Affairs, November 20, 2000, https://mfa.gov.il/MFA /IsraelExperience/History/Pages/Jerusalem%20-%20The%20Upper%20 City%20during%20the%20Second%20Templ.aspx?ViewMode=Print.

15.　Elliot Spagat, "US Identifies 3,900 Children Separated at Border Under Trump," Associated Press, June 8, 2021, https://apnews.com/article/az-state -wire-donald-trump-immigration-lifestyle-government-and-politics-54e2e5b bff270019d8bda3c81161c7c7.

16.　Angelo Fichera, "False Claim of Immigrant Children Deaths Under Obama," Factcheck.org, January 3, 2019, https://www.factcheck.org/2019/01/false -claim-of-immigrant-children-deaths-under-obama/.

17.　https://www.youtube.com/watch?v=JmaZR8E12bs.

18.　Masha Gessen, "Why the Tree of Life Shooter was Fixated on the Hebrew Immigrant Aid Society," *New Yorker*, October 27, 2018, https://www.new yorker.com/news/our-columnists/why-the-tree-of-life-shooter-was-fixated -on-the-hebrew-immigrant-aid-society.

19.　Dara Lind, "HIAS, the Jewish Refugee-Aid Group Targeted by the Pittsburgh Synagogue Shooter, Explained by Its President," Vox, October 29, 2018, https://www.vox.com/2015/9/25/9392151/hias-jewish-refugees-im migrants.

20.　Brett Samuels, "Trump: 'I Wouldn't Be Surprised' if Soros Were Paying for Migrant Caravan," *Hill*, October 31, 2018, https://thehill.com/homenews /administration/414171-trump-i-wouldnt-be-surprised-if-soros-were-paying -for-migrant-caravan.

21.　Linda Qiu, "Did Democrats, or George Soros, Fund Migrant Caravans? Despite Republican Claims, No," *New York Times*, October 20, 2018, https:// www.nytimes.com/2018/10/20/world/americas/migrant-caravan-video -trump.html.

22.　Jennifer, phone interview, August 13, 2021.

23.　Hannah Lebovits, phone interview, March 2, 2021.

24.　"Chief Rabbi Says It Doesn't Matter if Pittsburgh's Tree of Life Is a Synagogue," *Times of Israel*, October 29, 2018, https://www.timesofisrael.com /israels-chief-rabbi-wont-call-pittsburghs-tree-of-life-a-synagogue/.

25.　See, for example, Stephen Lurie, "The Dismal Failure of Jewish Groups to

Confront Trump," *New Republic*, October 24, 2017, https://newrepublic.com/article/145482/dismal-failure-jewish-groups-confront-trump.

26. Daniel Elbaum, Zoom interview, September 21, 2020.

27. Alex Kane, "How the ADL's Israel Advocacy Undermines Its Civil Rights Work," *Jewish Currents*, February 8, 2021, https://jewishcurrents.org/how-the-adls-israel-advocacy-undermines-its-civil-rights-work/.

28. https://twitter.com/MortonAKlein7/status/1269321723763200013.

29. Yomat Shimron, "In Voting, Orthodox Jews Are Looking More Like Evangelicals," Religion News Service, February 19, 2021, https://religionnews.com/2021/02/19/the-political-chasm-between-left-and-right-is-tearing-orthodox-jews-apart/.

30. *Priorities of Trump Voters vs. Biden Voters in the Orthodox Jewish Community: A Post-Election Analysis*, Nishma Research, November 17, 2020, https://www.jewishdatabank.org/content/upload/bjdb/2020_Trump_Voters,_Biden_Voters_Orthodox_Jews_Post-Election_Survey_Summary_Report.pdf.

31. Ron Kampeas, "After Afghanistan's Last Jew Refused to Leave, His Would-Be Jewish Rescuers Helped Dozens of Other Afghans Escape Instead," Jewish Telegraphic Agency, August 27, 2021, https://www.jta.org/2021/08/27/global/after-afghanistans-last-jew-refused-to-leave-his-would-be-jewish-rescuers-helped-dozens-of-other-afghans-escape-instead.

32. https://twitter.com/ballabon/status/1204660696912404480.

33. Judy Maltz, "Israel Ignores High Court Ruling on non-Orthodox Conversions," *Haaretz*, December 6, 2021, https://www.haaretz.com/israel-news/.premium-israeli-interior-ministry-ignores-high-court-ruling-on-conversions-1.10443541.

34. Barak Dunayer, phone interview, August 27, 2021.

35. https://twitter.com/RJC/status/1163900725262540805.

36. Dave Goldiner, "Rudy Giuliani Says Jewish Democrats ARE Disloyal to Israel for Backing Tlaib and Omar," *New York Daily News*, August 21, 2019, https://www.nydailynews.com/news/politics/ny-rudy-giuliani-trump-jews-dual-loyalty-adl-20190821-vn6fplkvfrge3ioert6tad6ki4-story.html.

37. Jake Turx, "EXCLUSIVE: Donald J. Trump's Winter White House // Catching Up with the 45th President Ahead of His Departure from Mar-a-Lago," *Ami*, June 16, 2021, https://www.amimagazine.org/2021/06/16/donald-j-trumps-winter-white-house/.

38. *Jewish Americans in 2020*, Pew Research Center, May 11, 2021, https://www.pewforum.org/2021/05/11/jewish-americans-in-2020/.

39. Emily Tamkin, *The Influence of Soros: Politics, Power, and the Struggle for an Open Society* (New York: Harper, an imprint of HarperCollins, 2020), 237.

40. "Yair Netanyahu: Soros' Organizations Destroying Israel from the Inside," *Hungary Today*, October 30, 2019, https://hungarytoday.hu/yair-netanyahu-soros-organizations-destroying-israel-from-the-inside/.

41. Olivia Nuzzi, "A Conversation with Rudy Giuliani over Bloody Marys at the Mark Hotel," *New York Magazine*, December 23, 2019, https://nymag.com/intelligencer/2019/12/a-conversation-with-rudy-giuliani-over-bloody-marys.html.

42. *Jewish Americans in 2020*, Pew Research Center, May 11, 2021, https://www
.pewforum.org/2021/05/11/jewish-practices-and-customs/.

43. Keren McGinity, *Still Jewish: A History of Women and Intermarriage in America*
(New York: New York University Press, 2009), 6.

44. https://twitter.com/JasonSCampbell/status/1393317713485905920.

45. Eli Valley, phone interview, October 14, 2021.

46. Steven Pruzansky, "The Consequences of Intermarriage," *Jerusalem Post*,
July 24, 2021, https://www.jpost.com/opinion/the-consequences-of-inter
marriage-opinion-674783.

47. Daniel Tamkin, phone interview, August 8, 2020.

48. Name withheld, phone interview, August 11, 2021.

49. Jennifer, phone interview.

50. Name withheld, phone interview, July 25, 2021.

51. Shoshana, phone interview, September 6, 2021.

52. Rebecca Pierce, "Black Jewish Voices Are Finally Being Heard. So Is the
Racist Backlash.," *Forward*, January 24, 2019, https://forward.com/opinion
/418143/black-jewish-voices-are-finally-being-heard-so-is-the-racist-back
lash/.

53. Max Daniller-Varghese, phone interview, September 1, 2021.

54. Jamie Yong, phone interview, August 5, 2021.

55. Arno Rosenfeld, "Largest Study Ever of Jews of Color Reports Wide-
spread Discrimination," *Forward*, August 12, 2021, https://forward.com
/news/474074/jews-of-color-study-discrimination-black-asian-latinx/.

56. Ira M. Sheskin and Arnold Dashefsky, "How Many Jews of Color Are
There?," eJewish Philanthropy, May 17, 2020, https://ejewishphilanthropy
.com/how-many-jews-of-color-are-there/.

57. See, for example, https://twitter.com/scarabbi/status/1426956741405777
922, and https://twitter.com/DavidJFryman/status/1427065629849366531.

Chapter 9: Pushing Jews

1. Max Fisher, "Here Is Clinton and Sanders's Remarkable Exchange on Israel-
Palestine—and Why It Matters," Vox, April 15, 2016, https://www.vox.com
/2016/4/15/11437602/clinton-sanders-israel-palestine-debate.

2. "Wright Says 'Jews' Keeping Him from Obama," Associated Press, June 11,
2009, https://www.nbcnews.com/id/wbna31246353.

3. See, for example, Samuel Gordon, "Obama and the Jews: An Inside Perspective," April 11, 2013, Shalom Hartman Institute, https://www.hartman.org
.il/obama-and-the-jews-an-inside-perspective/.

4. See, for example, "AIPAC Pummels Obama," *Washington Post*, March 2,
2015, https://www.washingtonpost.com/opinions/aipac-pummels-obama
/2015/03/02/e05ee4c2-c10d-11e4-9ec2-b418f57a4a99_story.html.

5. Emily Tamkin and Alexis Levinson, "Israel Will Be the Great Foreign Policy Debate of the Democratic Primary," BuzzFeed News, January 16, 2019,
https://www.buzzfeednews.com/article/emilytamkin/israel-dnc-bds
-democrats-2020.

6. "Alisa Biran, Jeremy Ben-Ami," *New York Times*, February 18, 2001, https://www.nytimes.com/2001/02/18/style/alisa-biran-jeremy-ben-ami.html.

7. https://jstreet.org/about-us/staff/jeremy-ben-ami/#.YXWnKtnMI U.

8. Emily Tamkin, "J Street Is Minding the Mainstream," *New Republic*, October 31, 2019, https://newrepublic.com/article/155564/j-street-minding-mainstream.

9. See, for example, Abraham J. Edelheit, Review of "A New Voice for Israel: Fighting for the Survival of the Jewish Nation," Jewish Book Council, August 30, 2011, https://www.jewishbookcouncil.org/book/a-new-voice-for-israel-fighting-for-the-survival-of-the-jewish-nation.

10. Ethan Bronner, "US Group Stirs Debate on Being 'Pro-Israel,'" *New York Times*, March 24, 2011, https://www.nytimes.com/2011/03/25/world/middleeast/25israel.html.

11. David Friedman, "Read Peter Beinart and You'll Vote for Donald Trump," *Arutz Sheva*, June 5, 2016, https://www.israelnationalnews.com/Articles/Article.aspx/18828.

12. Peter Beinart, "I No Longer Believe in a Jewish State," *New York Times*, July 8, 2020, https://www.nytimes.com/2020/07/08/opinion/israel-annexation-two-state-solution.html.

13. Tamkin and Levinson, "Israel Will Be the Great Foreign Policy Debate of the Democratic Primary."

14. Elizabeth Podrebarac Sciupac and Gregory A. Smith, "How Religious Groups Voted in the Midterm Elections," Pew Research Center, November 7, 2018, https://www.pewresearch.org/fact-tank/2018/11/07/how-religious-groups-voted-in-the-midterm-elections/.

15. Laura E. Adkins, "By the Numbers: 3 Key Takeaways from the 2016 Jewish Vote," *Forward*, November 9, 2016, https://forward.com/news/353914/by-the-numbers-3-key-takeaways-from-the-2016-jewish-vote/.

16. Scott Clement, "Jewish Americans Support the Iran Nuclear Deal," *Washington Post*, July 27, 2015, https://www.washingtonpost.com/news/the-fix/wp/2015/07/27/jewish-americans-support-the-iran-nuclear-deal/.

17. Tamkin and Levinson, "Israel Will Be the Great Foreign Policy Debate of the Democratic Primary."

18. Saphora Smith, Lawahez Jabari, and Ayman Mohyeldin, "Israel Bars Muslim Reps. Omar and Tlaib from Visiting the Country," *NBC News*, NBC, August 15, 2019, https://www.nbcnews.com/news/world/israel-will-block-muslim-reps-omar-tlaib-visiting-country-deputy-n1042606.

19. Michael Kranish, "Eight Minutes with the President: Rep. Rashida Tlaib, the Lone Palestinian American in Congress, Gains Relevance in Israel Debate," *Washington Post*, May 20, 2021, https://www.washingtonpost.com/politics/eight-minutes-with-the-president-rep-rashida-tlaib-the-lone-palestinian-american-in-congress-gains-relevance-in-israel-debate/2021/05/20/fe3139c6-b8c5-11eb-bb84-6b92dedcd8ed_story.html.

20. "What the Jewish Left Learned from Occupy," *Jewish Currents*, October 5, 2021, https://jewishcurrents.org/what-the-jewish-left-learned-from-occupy.

21. Jameelah Nasheed, "36 Jewish Activists Were Arrested While Protesting an ICE Detention Center in New Jersey," *Teen Vogue*, July 1, 2019, https://www

.teenvogue.com/story/36-jewish-activists-arrested-protesting-ice-detention
-center-new-jersey.

22. Tae Phoenix, "Jews Against ICE: We're Doing What the Gentiles of Europe Should Have Done," *Newsweek*, July 3, 2019, https://www.news
week.com/jews-against-ice-doing-gentiles-should-have-done-concentration
-camps-1447386.

23. Carol Kuruvilla, "Jewish American Activists Observe the High Holidays with Protests Against ICE," HuffPost, October 4, 2019, https://www.huffpost
.com/entry/days-of-awe-jewish-protests_n_5d94b483e4b0e9e760560275.

24. Liel Leibovitz, "Us and Them," Tablet, May 25, 2021, https://www.tablet
mag.com/sections/israel-middle-east/articles/zionism-liel-leibovitz.

25. Justin Baragona, "Meghan McCain Attempts to Lecture Chuck Schumer About Antisemitism," Daily Beast, July 23, 2021, https://www.thedaily
beast.com/meghan-mccain-attempts-to-lecture-chuck-schumer-about-anti
semitism.

26. Emily Tamkin, "How Should US Antisemitism Be Defined in the Biden Era," *New Statesman*, February 4, 2021, https://www.newstatesman.com/world
/americas/north-america/2021/02/how-should-us-anti-semitism-be-defined
-biden-era-0.

27. "The Jerusalem Declaration on Antisemitism," https://jerusalemdeclar
ation.org/.

28. David Schraub, "A New Definition of Antisemitism Is Out, and the Antisemites Love It," *Haaretz*, April 7, 2021, https://www.haaretz.com/world-news
/.premium.HIGHLIGHT-a-new-definition-of-antisemitism-is-out-and-the
-antisemites-love-it-1.9685765.

29. Isaac Chotiner, "How Antisemitism Rises on the Left and Right," *New Yorker*, January 2, 2020, https://www.newyorker.com/news/q-and-a/how
-anti-semitism-rises-on-the-left-and-right.

30. Adam Nagourney, "In U.C.L.A. Debate over Jewish Student, Echoes on Campus of Old Biases," *New York Times*, March 5, 2015, https://www.ny
times.com/2015/03/06/us/debate-on-a-jewish-student-at-ucla.html.

31. Ruth Graham and Liam Stack, "U.S. Faces Outbreak of Anti-Semitic Threats and Violence," *New York Times*, May 26, 2021, https://www.nytimes
.com/2021/05/26/us/anti-semitism-attacks-violence.html.

32. Dimi Reider, "Israel-Palestine: Why Netanyahu and Hamas Both Risk Losing Control of the Conflict," *New Statesman*, May 13, 2021, https://www
.newstatesman.com/world/2021/05/israel-palestine-why-netanyahu-and
-hamas-both-risk-losing-control-conflict.

33. Adam Serwer, *The Cruelty Is the Point: The Past, Present, and Future of Trump's America* (New York: One World, 2021), 165–77.

34. Mike DeBonis and Rachael Bade, "Rep. Omar Apologizes After House Democratic Leadership Condemns Her Comments as 'Anti-Semitic Tropes,'" *Washington Post*, February 11, 2019, https://www.washingtonpost.com
/nation/2019/02/11/its-all-about-benjamins-baby-ilhan-omar-again-accused
-anti-semitism-over-tweets/.

35. Jeremy Slevin, phone interview, October 7, 2021.

36. "Rep. Omar Introduces The MBS MBS Act to Sanction Saudi Arabian Crown

Prince Mohammed bin Salman," Ilhan Omar Serving the 5th District of Minnesota, March 2, 2021, https://omar.house.gov/media/press-releases/rep-omar-introduces-mbs-mbs-act-sanction-saudi-arabian-crown-prince-mohammed.

37. "Rep. Ilhan Omar Leads Letter Calling for Accountability and Deescalation in Kashmir," Ilhan Omar Serving the 5th District of Minnesota, September 17, 2019, https://omar.house.gov/media/press-releases/rep-ilhan-omar-leads-letter-calling-accountability-and-deescalation-kashmir.

38. "Rep. Omar Leads Letter to CEOs, Including Apple, Amazon, and Google, Condemning the Use of Forced Uyghur Labor in China," Ilhan Omar Serving the 5th District of Minnesota, April 6, 2020, https://omar.house.gov/media/press-releases/rep-omar-leads-letter-ceos-including-apple-amazon-and-google-condemning-use.

39. "Rep. Ilhan Omar Statement on the Situation in Ethiopia," Ilhan Omar Serving the 5th District of Minnesota, February 19, 2021, https://omar.house.gov/media/press-releases/rep-ilhan-omar-statement-situation-ethiopia.

40. Slevin, phone interview.

41. See, for example, "Jewish Leaders Condemn Tlaib's 'Antisemitic Dog Whistle' in Recent Comments," *Times of Israel*, August 7, 2021, https://www.timesofisrael.com/jewish-leaders-condemn-tlaibs-antisemitic-dog-whistle-in-recent-comments/.

42. "A Threshold Crossed: Israeli Authorities and the Crimes of Apartheid and Persecution," Human Rights Watch, April 27, 2021, https://www.hrw.org/report/2021/04/27/threshold-crossed/israeli-authorities-and-crimes-apartheid-and-persecution#.

43. "A Regime of Jewish Supremacy from the Jordan River to the Mediterranean Sea: This Is Apartheid," B'Tselem, January 12, 2021, https://www.btselem.org/publications/fulltext/202101_this_is_apartheid.

44. See, for example, Gideon Levy and Alex Levac, "The Illegal Settler Outpost Has Running Water. Its Palestinian Neighbors Don't. This Is Apartheid at Its Starkest," *Haaretz*, September 24, 2021, https://www.haaretz.com/israel-news/twilight-zone/MAGAZINE-the-settler-outpost-has-water-its-palestinian-neighbors-don-t-this-is-apartheid-1.10237053.

45. Philissa Cramer, "Dozens of US Rabbinical Students Sign Letter Calling for American Jews to Hold Israel Accountable for Its Human Rights Abuses," Jewish Telegraphic Agency, May 14, 2021, https://www.jta.org/2021/05/14/united-states/dozens-of-us-rabbinical-students-sign-letter-calling-for-american-jews-to-hold-israel-accountable-for-its-human-rights-abuses.

46. Scott Goodstein, phone interview, December 22, 2020.

47. Eric Cortellessa, "Democratic Hopeful Sanders Urges Giving Chunk of US Military Aid to Gaza Instead," *Times of Israel*, October 28, 2019, https://www.timesofisrael.com/democratic-hopeful-sanders-urges-giving-chunk-of-us-military-aid-to-gaza-instead/.

48. Tamkin, "J Street Is Minding the Mainstream."

49. Ali Harb, "Bernie Sanders Says He Will Skip AIPAC Conference," Middle East Eye, February 24, 2020, https://www.middleeasteye.net/news/bernie-sanders-says-he-will-skip-aipac-conference.

50. https://twitter.com/aipac/status/1234239804734394368?lang=en, and https://twitter.com/aipac/status/1233958882965368832.

51. Eric Cortellessa, "Bloomberg to AIPAC: I'll Never Condition Aid to Israel, No Matter Who's PM," *Times of Israel*, March 2, 2020, https://www.timesof israel.com/bloomberg-to-aipac-ill-never-condition-aid-to-israel-no-matter -whos-pm/.

52. Aiden Pink, "Sanders at Debate: I'm Proud of Being Jewish, but Netanyahu a 'Reactionary Racist,'" *Forward*, February 25, 2020, https://forward.com /fast-forward/440461/sanders-bloomberg-israel-debate/.

53. https://twitter.com/MikeBloomberg/status/1232496557649350657.

54. See, for example, Tiana Lowe, "Bernie Sanders Has an Antisemitism Problem," *Washington Examiner*, December 13, 2019, https://www.washingtonexam iner.com/opinion/bernie-sanders-campaign-has-an-anti-semitism-problem.

55. Ron Kampeas, "Bernie Sanders: Being Jewish Is One of Two Factors That Shaped His Outlook," *Times of Israel*, February 7, 2020, https://www.times ofisrael.com/bernie-sanders-being-jewish-is-one-of-two-factors-that-shaped -his-outlook/.

56. Bernie Sanders, "How to Fight Antisemitism," *Jewish Currents*, November 11, 2019, https://jewishcurrents.org/how-to-fight-antisemitism.

57. See, for example, Talia Lavin, "To Dream of a Jewish President," *New Republic*, February 13, 2020, https://newrepublic.com/article/156552/dream -jewish-president.

58. https://twitter.com/IfNotNowOrg/status/1247940274736762881.

59. https://twitter.com/jewsforbernie/status/1247912924930281472.

60. Sam Stein, "Biden Embraces Endorsement of Liberal Jewish Group J Street," Daily Beast, April 17, 2020, https://www.thedailybeast.com/joe -biden-embraces-endorsement-of-liberal-jewish-group-j-street.

61. Tamkin, "J Street Is Minding the Mainstream."

62. Mallory Simon, "Not Just Neo-Nazis with Tiki Torches: Why Jewish Students Say They Also Fear Cloaked Antisemitism," CNN, July 1, 2021, https://www .cnn.com/2021/06/30/us/american-anti-semitism-students-soh/index .html.

63. Michelle Boorstein, "New Poll: Young US Jews Becoming More Orthodox As American Judaism Splits Between Devout and Secular," *Washington Post*, May 11, 2021, https://www.washingtonpost.com/religion/2021/05/11 /orthodox-jews-poll-secular-trump-republican/.

64. Dov Lipman, phone interview, October 4, 2021.

65. Aaron Freedman, "What Happened to IfNotNow?," *Jewish Currents*, April 26, 2021, https://jewishcurrents.org/what-happened-to-ifnotnow.

66. Klil H. Neori, phone interview, July 25, 2021.

67. Yousef Munayyer, phone interview, October 13, 2021.

68. Jacob Plitman, phone interview, March 4, 2021.

69. See, for example, Joshua Leifer, "A Threat, Not an Offer," *Jewish Currents*, January 29, 2020, https://jewishcurrents.org/a-threat-not-an-offer/.

70. Asad Dandia, "When 'Jewish Security' Means Muslim Surveillance," *Jewish Currents*, February 24, 2020, https://jewishcurrents.org/when-jewish -security-means-muslim-surveillance.

71. Nylah Burton, "The *Forward's* 'Both Sides' Approach Has Failed," *Jewish Currents*, May 15, 2019, https://jewishcurrents.org/the-forward-s-both-sides-approach-has-failed.

72. Mari Cohen, "They Want to Kick Us Out of This Land," *Jewish Currents*, October 15, 2021, https://jewishcurrents.org/they-want-to-kick-us-out-of-this-land.

73. Kaleem Hawa, "The Nakba Demands Justice," *Jewish Currents*, May 14, 2021, https://jewishcurrents.org/the-nakba-demands-justice.

74. Claire Schwartz, "For the Sake of Truth," *Jewish Currents*, October 21, 2021, https://jewishcurrents.org/for-the-sake-of-truth.

75. Plitman, phone interview.

76. Hannah Kahn, phone interview, August 25, 2021.

77. Elad Strohmayer, Zoom interview, January 14, 2021.

78. Jesse Chase-Lubitz, phone interview, February 10, 2021.

79. Logan Bayroff, phone interview, September 14, 2020.

80. *Jewish Americans in 2020*, Pew Research Center, May 11, 2021, https://www.pewforum.org/2021/05/11/jewish-americans-in-2020/.

81. See, for example, Hagar Shezaf and Jonathan Lis, "Israel Advances 1,300 Homes in West Bank Settlement in First Since Biden Sworn In," *Haaretz*, October 24, 2021, https://www.haaretz.com/israel-news/.premium-israel-advances-1–300-housing-units-for-jews-in-west-bank-in-first-since-biden-sworn-1.10319962.

82. *Times of Israel* staff, "Bennett Says He Won't Meet Mahmoud Abbas, Palestinian State a 'Terrible Mistake,'" *Times of Israel*, September 14, 2021, https://www.timesofisrael.com/bennett-says-he-wont-meet-mahmoud-abbas-palestinian-state-a-terrible-mistake/.

83. Name withheld, personal interview, Tel Aviv, October 7, 2021.

84. Plitman, phone interview.

85. Jordan Fraade, Zoom interview, August 19, 2021.

86. Jaz Twersky, phone interview, August 27, 2021.

87. Meli Sameh, Zoom interview, August 17, 2021.

88. Mordechai Lightstone, phone interview, July 21, 2021.

89. Ike Swetlitz, phone interview, July 25, 2021.

90. https://twitter.com/TheRaDR/status/1392952511603093509.

91. Evan Mintz, phone interview, October 25, 2021.

92. Shuja Haider, "Eli Valley Is Not Sorry," Popula, March 11, 2019, https://popula.com/2019/03/11/eli-valley-is-not-sorry/.

93. Rachel Gross, *Beyond the Synagogue: Jewish Nostalgia as Religious Practice* (New York: New York University Press, 2021), 200.

94. Anthony Russell, phone interview, October 26, 2021.

95. Andrew Silow-Carroll, "Times Are Shvakh, but Yiddish Classes Are Booming," *Jewish Week, Times of Israel*, August 7, 2020, https://jewishweek.timesofisrael.com/times-are-shvakh-but-yiddish-classes-are-booming/.

96. Mikhl Yashinsky, phone interview, September 25, 2020.

97. Rabbi Angela Buchdahl, "We Are Family: Rethinking Race in the Jewish Community," Yom Kippur 5781/2020, September 28, 2020, https://www.youtube.com/watch?v=FNhG8aW6gbI&t=1020s.

98. Christian Todd, "A Queer, Black, Female Rabbi's Fight for Racial Equality in Judaism," *ABC 7 Eyewitness News*, ABC, February 24, 2021, https://abc7 chicago.com/localish-black-lives-matter-racial-equity/10364568/.

99. Sandra Lawson, phone interview, October 18, 2021.

100. Russell, phone interview.

101. "Yiddish, Anti-Racist Practice, and the Transformation of Jewish Communities," Museum of Jewish Heritage, August 27, 2020, https://mjhnyc.org /blog/yiddish-anti-racist-practice-and-the-transformation-of-jewish-com munities/.

102. Shoshana, phone interview, September 6, 2021.

103. Hanah Bloom, "Anti-Asian Violence Has Been Ignored for Too Long. Here's How Jews Can Help.," Hey Alma, March 18, 2021, https://www.heyalma .com/anti-asian-violence-has-been-ignored-for-too-long-heres-how-jews-can -help/.

104. Jessy Kuehne, "I Converted Through Reform Judaism. Stop Telling Me I'm Not a Jew.," Hey Alma, October 26, 2020, https://www.heyalma.com/i-con verted-through-reform-judaism-stop-telling-me-im-not-a-jew/.

105. Elysse DaVega, "The Jewish Internet Helped Me Claim My Sephardic Jewish Heritage," Hey Alma, April 6, 2021, https://www.heyalma.com/the-jewish internet-helped-me-claim-my-sephardic-jewish-heritage/.

106. Molly Tolsky, phone interview, September 23, 2020.

107. Devin E. Naar, "A Century Ago, Jewish Salonica Burned," *Times of Israel*, August 18, 2017, https://www.timesofisrael.com/a-century-ago-jewish-sal onica-burned/.

108. Devin Naar, phone interview, October 1, 2020.

109. Tomer Shani, Zoom interview, September 1, 2021.

Conclusion

1. Daniel Zemel, personal interview, Washington, DC, October 14, 2021.

2. Cindy Cardinal, phone interview, October 9, 2021.

INDEX

Abrams, Elliott, 133, 135
Abzug, Bella, 100
Acosta, Jim, 193–194
Adelson, Miriam, 160–161
Adelson, Sheldon, 160, 205
Adler, Cyrus, 24
affirmative action, 103–104
Agudath ha-Rabbanim (Union of Orthodox Rabbis), 19
AJC. *See* American Jewish Committee
al-Adraa, Basil, 235
Aleichem, Sholem, 72
Alexeyeva, Ludmilla, 174
aliyah, 81–82, 231–232
Amateau, Albert, 22
Ambivalent Embrace (Kranson), 62, 72
American Board of Rabbis, 182
American Council for Judaism, 89
American Federation of Labor, 55
American Israel Public Affairs Committee (AIPAC), 200, 217, 218, 221, 228, 229
Americanization. *See* assimilation
American Jewish Committee (AJC)
 civil rights movement and, 102, 104
 founding of, 23–24
 Jewish relationship to whiteness and, 24

McCarthyism and, 145
neoconservative Jews and, 130
responses to Nazi Germany, 45–46
Rosenberg case and, 68–69
Soviet Jews and, 175
Trump administration and, 200–201
Zionism and, 82, 83
See also Commentary
American Jewish Congress, 45–46, 83, 99, 101, 104, 166
American Jewish Philanthropic Complex, The: The History of a Multibillion-Dollar Institution (Corwin Berman), 153
American Judaism: A History (Sarna), 30, 39
Amini, Esther, 179
Ammud, 240
anticommunism. *See* Cold War; McCarthyism
Anti-Defamation League (ADL)
 Christian right and, 138
 civil rights movement and, 102, 104
 founding of, 23
 Rosenberg case and, 68
 Trump administration and, 3, 201
Antin, Mary, 25–26

Index

ABOUT THE AUTHOR

EMILY TAMKIN is the author of *The Influence of Soros* and is the senior editor, US, of the *New Statesman*. Her work has appeared in the *Columbia Journalism Review*, the *Economist*, the *New Republic*, *Politico*, *Slate*, and the *Washington Post*, among other publications. She previously covered foreign affairs on staff at *Foreign Policy* and BuzzFeed News. She studied Russian literature and culture at Columbia University and Russian and East European studies at the University of Oxford. She has conducted research on Soviet dissidence on a Fulbright Fellowship in Germany. She was also a Council on Foreign Relations International Affairs fellow in New Delhi, India. She lives in Washington, DC.